(Great) Grampie Fage
Xmas 1995.

Here is a little history of the area where we live hope you enjoy it. Merry Christmas

Paul Cindi
Kiersten Eilish + ½.

THE GOLDEN MILE

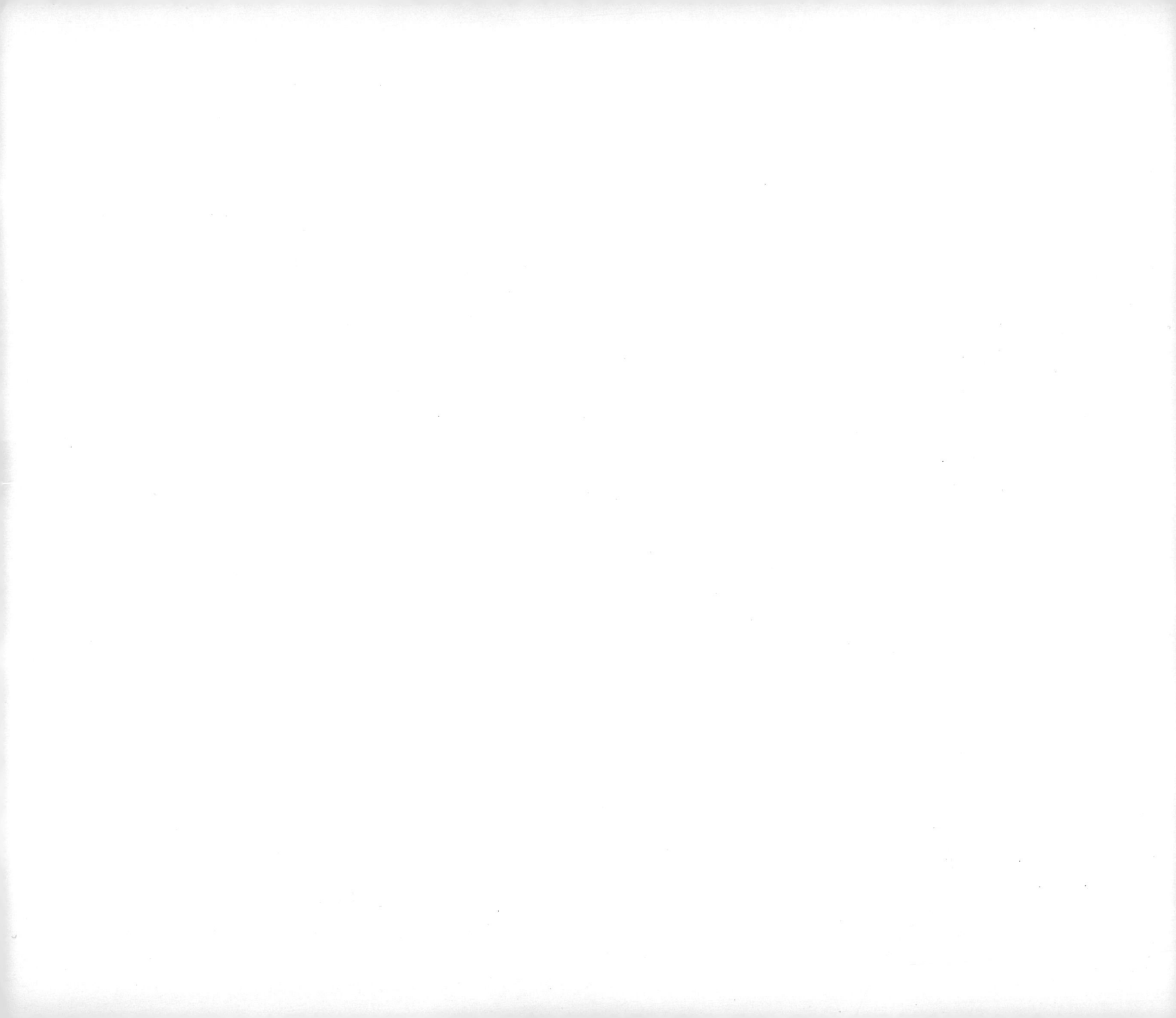

THE GOLDEN MILE

GEOFFREY BLAINEY

A Rathdowne Book
ALLEN & UNWIN

© Geoffrey Blainey 1993

This book is copyrighted under the Berne Convention. No reproduction without permission. All rights reserved.

First published in 1993

A Rathdowne Book
Allen & Unwin Pty Ltd
9 Atchison Street, St Leonards NSW 2065, Australia

National Library of Australia
cataloguing-in-publication data:

Blainey, Geoffrey, 1930-
The golden mile.

Includes index.
ISBN 1 86373 480 5.

1. Gold mines and mining — Western Australia — Kalgoorlie Region — History. II. Kalgoorlie Region (W.A.) — History. I. Title.

994.16

Designed by Guy Mirabella
Set in Berkeley Old Style by Bookset, Melbourne
Printed by Kim Hup Lee, Singapore

Acknowledgements

These members of the Eastern Regional Council of the Chamber of Mines and Energy of Western Australia have sponsored the writing of this book: Ashton Mining Ltd, Central Kalgoorlie Gold Mines, Central Norseman Gold Corporation Ltd, Coolgardie Gold NL, Dominion Mining Ltd, Eltin Ltd, Forsayth Mining Services Ltd, Herald Resources Ltd, Jubilee Mine—PosGold (GMK), Kalgoorlie Consolidated Gold Mines Pty Ltd, Metall Mining Australia Pty Ltd, Mt McClure Gold Mine, Mt Martin Gold Mines NL, Mt Monger Gold Project, Newcrest Mining Ltd, Pancontinental Goldmining Operations, Placer (Granny Smith) Pty Ltd, Poseidon Gold Ltd—Kaltails Project, Sons of Gwalia Ltd, Spargos Mining NL, Sundowner Minerals NL and Western Mining Corporation Ltd.

CONTENTS

Preface vii
1. The Unlikely One 1
2. Is That Gold? 6
3. Unlocking the Jewel Chambers 23
4. The Torment of the Water King 55
5. Along the Crowded Mile 77
6. War and the 'Foreigners' 101
7. In the Path of the Whirlwind 113
8. The Decade was Upside Down 128
9. Sole Survivor 144
10. The Golden Hole 159

Sources of Information 173
Index 177

PREFACE

THIS BOOK WAS WRITTEN for the centenary of Kalgoorlie — the most productive goldfield in one of the world's great gold nations. Kalgoorlie is a rarity: a mining field worked continuously for 100 years. At times it was close to extinction. In grave trouble in 1929, it was saved by the world depression which curiously made it the most prosperous town in a depressed country. Again in 1975 it was within a few days of closing down. The slumps were punctuated by booms when the gold pouring out of the Golden Mile helped to shape the history of Western Australia.

This is primarily a history of the mines and not of the towns above the ground. While many books and booklets have been written about people, periods and episodes in the history of Kalgoorlie, I think this is the first book solely on the history of its many mines. It was designed as a short history.

The idea of writing the book came from the Chamber of Mines and Energy of Western Australia, which was founded in Coolgardie, spent most of its years in Kalgoorlie when that was the capital city of Western Australian mining, and is now based in Perth. I thank its chief executive officer, Peter Ellery, and its regional secretary, Gary Arcus, for their co-operation at every step. I also thank David Blight of Mt Martin who initially acted as liaison officer on the goldfields.

People with strong goldfields links gave reminiscences or guided me to sources. They included Reg Buckett, Ian Burston, Graeme Campbell, Denis Cumming, Darren Head, Jack Manners, Ron Manners, Brian Phillips, John Rowe, Professor Eric Rudd and Peter Wreford. Martyn and Audrey Webb, whose research on the eastern goldfields in recent years is voluminous, did not hesitate to help though their own book will appear in 1993. At the fine Museum of the Goldfields in Kalgoorlie the curator, Pam Moore, gave me access to many of the photographs, and Joanna Sassoon of the Battye Library in Perth found others. Terry Kings-Lynne of the WA Water Authority sent me the crucial report by C. Y. O'Connor on the goldfields water scheme.

In Melbourne, Jillian Coster and Ben Scott spent time seeking information in early newspapers and parliamentary reports. Tom and Noelene Scally of the Quality Plaza Inn in Kalgoorlie went out of their way to help. Staff of the La Trobe Library, the State Records of South Australia, the Mortlock Library of South Australiana, the history room of the Ballarat Regional Library, the Battye Library and the Western Australian State Archives were helpful. I thank them all.

Various chapters in the draft manuscript were read, at short notice, by Keith Quartermaine of the Kalgoorlie School of Mines, by Sir Laurence Brodie-Hall in Perth, and by Gilbert Ralph of Western Mining Corporation in Melbourne. Experts on many facets of Kalgoorlie, their comments and criticisms were gratefully received.

GEOFFREY BLAINEY   *Melbourne, November 1992*

CHAPTER ONE

THE UNLIKELY ONE

PADDY HANNAN, the discoverer of Kalgoorlie, did not quite look like a hero. If all the people who did most to bring prosperity to Australia — the living and the dead — could be brought together and paraded on television, Hannan would probably be singled out as the one chosen in error. He was awkward in dress, bald, beady-eyed, and inclined to be apprehensive. Though definitely alert and intelligent, he was secretive. If interviewed, he might have paused for some time before he spoke, releasing each word slowly and courteously in a County Clare accent. At times he resembled a hermit, but that was not surprising. The middle years of his working life had been spent, alone or with a couple of other men, in the hot outback where he followed an occupation in which secrecy was vital.

The momentous month in his life was June 1893. He was then about 50 — the exact date of his birth is not known. One of the most experienced prospectors in those dry areas of Australia that were now the front line for the discovery of minerals, he was a son of the cold mist and rain. Hannan spent his childhood years near the Shannon River in Ireland, then worked as a gold miner in the wet mines of Ballarat in the 1860s and in the cold South Island of New Zealand for part of the 1870s before moving to a succession of rushes on the red soil of the Australian interior. In 1886 he was in the rush to arid Teetulpa (SA), between Port Augusta and Broken Hill. Three years later he crossed the Great Australian Bight in a steamship to Western Australia where he made his way from gold camp to gold camp, finally resolving to cross the plains to the new hessian town of Coolgardie in 1892. He set out soon after hearing that Bayley and Ford had found gold — an impressive feat of discovery, because the find lay about 120 barren miles beyond the nearest goldfield.

A few people who left Perth to travel to the Coolgardie gold rush went a short distance by train and then hired riding horses. Everyone else walked, paying a daily sum of money for the privilege — if they were lucky — of placing their swag on a dray or packhorse alongside which they tramped. Hannan walked. The journey normally took a fortnight, a footsore fortnight, of which the final seven or eight days were spent in tramping from the last town, Southern Cross, to the new rush.

When Hannan arrived, Coolgardie resembled the end of the earth. Its one attraction was the rich patches of gold within a few inches of the surface. The gold offered pleasing but not princely rewards to hard-working, sharp-eyed gold diggers. Hannan must have found gold or he could not have survived in a town with probably the highest cost of living in Australia.

By the winter of 1893, Coolgardie consisted of scattered clumps of tents and huts and bough-roofed sheds. There was no bank, no school, and in some weeks there were no women. No telegraph line came from distant Perth and no horse-drawn mail coach: the first bag of letters arrived by bicycle. There was not even much gold for the thousand or so men who spent each day searching for it, and most prospectors had both ears cocked for rumours of new finds of gold further out. The belief was strong that other goldfields would be found. Beyond the low hills in the distance and beyond the dazzling white salt-lakes, there must be other Coolgardies.

Suddenly a rush, spurred by rumour, headed for the north-east. Hannan prepared to join it. Teaming up with two other Irish prospectors named Flanagan and Shea, he located horses in the bush. Whether he just found the horses, hired or borrowed them is not known. Horses were a wonderful asset in this harsh country even though food and water for livestock were scarce in many months of the year.

The three men, at the tail end of the procession of gold seekers heading out, halted near a low hill about 25 miles from Coolgardie. There Hannan and Flanagan, near the hill now called Mt Charlotte, found two colours of gold. The date was 10 June 1893. Walking with heads down, the three men inspected a wide area of ground. They found more and more small pieces of gold in gravel and the dry red dust. Gold they also found embedded in white quartz rock. They had discovered the field later called Kalgoorlie.

Now and then prospectors passed by on their way to the distant goldfield which, it turned out, did not exist. While they were passing, Hannan and his mates almost certainly pretended that they were just camping, just resting. The hallmark of the professional prospector on a virgin goldfield, with gold waiting just below the surface for the first men who came along, was to remain silent. Shrewd Hannan and his mates were masters of silence. Ironically some prospectors camped nearby and spent the day playing the card game called 'nap'. They did not realise that Australia's richest goldfield had just been discovered within sight of their game and that the newfound gold was concealed in the camps of the three Irishmen.

The site of the first discovery can still be seen. It stands at the eastern end of what became the main street of Hannan's, as Kalgoorlie was first called. Four years later Hannan himself returned to the site and watched the planting of a peppercorn tree to mark the place where this wonderful goldfield was born.

For eight days the Irishmen had the new field to themselves. Working hard, they found a useful but not a sensational amount of gold. For each man the earnings were perhaps equal to a couple of years' wages. Then Hannan, clearly the leader of the threesome, decided that it would be wise to report the discovery to the officials. So he rode back to Coolgardie carrying several quarts of water to drink on the way and countless pieces of gold weighing about eight pounds. Excitedly but reluctantly he announced his find. He had no alternative to breaking the news if he was to claim the acres which, under Western Australia's mining law, were granted to the discoverer of a distinctive goldfield.

Next day Coolgardie saw a stampede of men, on horseback or foot, and the slow file of wagons and carts set out. In the space of a week Kalgoorlie held more men than Coolgardie. Those approaching the new goldfield could now recognise it from afar. Clouds of red dust hovered in the blue winter sky, the dust rising due to the method of dry-blowing which separated the light grains of worthless soil from the heavy grains of gold. As water was scarce, all men were dry-blowers. In effect they shovelled the gold-bearing dust into the air and allowed the barren dust to blow away, or they built a simple wooden contraption on tall legs and rocked it so that the lighter dust was parted from the heavy pieces of gold. It was easy to pick out the early Kalgoorlie miners — their faces were reddish brown.

Muscle was needed to mine parts of the goldfield. In Paddy Hannan's own ground, a few feet below the surface, very small nuggets of gold were being chipped from a strata of rock rather like cement. Miners using a pick or a hammer and gad could force the nuggets out of the white cement. Close by were two Irishmen, maybe Flanagan and Shea, who gathered a dazzling collection of small gold nuggets. They invited W. H. Corbould, a young mining engineer, to inspect their gold, and in the privacy of their camp they revealed small bags and tins from which they tipped hundreds of pieces of gold into three gold-panning dishes, there to be admired. Corbould wrote that the gold came in many shapes, some like a miniature version of 'the palm of one's hand' and some resembling little knucklebones. Judging by his description, the gold would have been valuable enough to buy a farm of liveable size in Victoria. The two Irishmen were generous, inviting Corbould to select a few nuggets. An honourable man, he chose just one — 'about half the size of a marble'.

Hardly any of those who camped at Kalgoorlie during its first two months could glimpse that it might be a very valuable field. On early indications it was not as precious as, say, 500 other goldfields found in eastern Australia in the last 40 years. Australia was sprinkled with goldfields, hundreds of them, that produced more gold than Kalgoorlie yielded in its first months. This newest field seemed to have little more than a pleasing sprinkling of gold that had eroded, over millions of years, from the hard rock. Most of the diggers at early Kalgoorlie had seen richer fields somewhere else.

Kalgoorlie provided one special consolation at this stage. Found during a severe depression, just one month after the worst banking crashes in Australian history, it offered a living to people who had lost their jobs. To a host of families hurt by the depression, Kalgoorlie would be a windfall.

High obstacles were in the way of those living in eastern Australia who thought of setting out for Hannan's find. The ship's fare from Melbourne to Western Australia was dear. Furthermore, the new gold rush was far from the coast. It was probably the first important goldfield in Australia to lie more than 300 miles inland. And it had other disadvantages. The summer was like an overheated

THE GOLDEN MILE

An early view of the Great Boulder hill, taken perhaps two years after Brookman's find. Not often is water seen in an early photograph of Kalgoorlie.

oven. The dry surrounding land could provide virtually no food for the miners. So the cost of living was high, with all the flour, sugar, tinned pudding, potatoes, shovels, picks and other supplies having to be carried on slow horse-drawn vehicles or strings of camels from the nearest railway station, not far from Perth. Above all, the new goldfield would lack water in many months. A large-scale goldfield could not depend on dry-blowing. It needed large quantities of water to treat a big tonnage of ore. Hitherto no permanent gold town of any size had been established in an area with such a low, capricious rainfall and such a high rate of evaporation.

The start of the first summer at Kalgoorlie brought heat waves. In daylight there was no refuge except tents and bough shelters: no shade worthy of the name. Cold water and ice could not be obtained. Water to quench the thirst — it was warm and brown — was absurdly expensive. In drought a gold miner needed perhaps a quarter ounce of gold just to pay for the week's drinking water. A few miles away from Hannan's find was a lake of salt water, and on the shores a few primitive condensing machines used wood stoves to boil the water and remove most of the salt, making the brackish water more fit for human consumption.

As Kalgoorlie's first summer set in, water became scarce and dearer. The surface gold was also becoming scarce. Many prospectors decided to go to the coast. Even riding a horse to Coolgardie, merely the first stage of the journey, could be risky; and on the long stretches of parched track towards Southern Cross long queues of horse teams stood at the few waterholes.

The warden in charge of the new eastern goldfields was John Finnerty, and in February 1894 he decided that a crisis was on hand. Mining laws in nearly all parts of Australia decreed that a prospector must actually work his ground or forfeit it; but at Hannan's the drought was so acute that Finnerty gave men permission to desert the settlement for a month. From his office in Southern Cross he sent urgent messages to the government in Perth. On 7 February his telegram read: 'The scarcity of water is becoming alarming.' Shrunken Kalgoorlie now held more people than the water could serve. For part of the summer few people remained on the parched field.

Paddy Hannan thought he would like to visit the sea and feel its cooling breezes. He had not seen the sea for five years so, leaving most of his possessions in his simple camp, he hurried away. Unknown to him, however, the goldfield was already entering a new era.

IS THAT GOLD?

CHAPTER TWO

THE RICHEST GROUND was taken not by Hannan but by men who arrived late. They were not even in the district when Hannan found the first gold. They were probably not even among the first 500 who walked over the ironstone hills which, unknown to all, concealed the real treasure.

Adelaide sent the two prospectors who, by normal standards, came too late. Then the third biggest city in Australia, Adelaide was not usually represented in strength at new gold rushes. It had long invested in mining shares but gold was not its traditional fancy. In the nineteenth century no other Australian colony was so rich as South Australia in copper mines but so poor in gold.

George Brookman, who in the 1880s had moved from selling groceries to selling shares in Adelaide's main street, King William Street, was attracted by gold. He first became interested in gold mines at the north Queensland field of Charters Towers, more than 2000 miles by steamship from Adelaide. Then the new Western Australian fields attracted him, and he formed a little syndicate to search for gold there. Eight South Australians and two Melbourne men each put up a few pounds each — enough to pay the fares for two prospectors to travel to Western Australia, equip themselves for the journey into the interior, and spend three or four months in searching. The capital of the syndicate was a mere 150 pounds.

It was crucial to select the right prospectors. Their first choice could hardly be called inspired. He was Brookman's young brother, Will, whose career had been in the city. As manager of a large jam and sauce factory in Adelaide, he virtually became bankrupt. Incompetence as well as bad luck seems to have caused his fall. To start life afresh he went to a minor goldfield near Dashwood's Gully, about 24 miles south of Adelaide. Sleeping in a tent at night he worked hard each day to win a miserable amount of gold. He was happy to leave.

Sam Pearce was selected as second man in the party. Nobody could complain about Sam, except that he didn't yet know much about gold. After emigrating as a toddler from England to Adelaide, Pearce had spent part of his childhood at the copper town of Kapunda where his father kept a general store. Young Sam then went to sea, returned to Kapunda and married the daughter of a copper miner. Eventually — at the age of 26 — he took up a wheat farm on the agricultural frontier. His farm was at Belalie, beyond Jamestown, and close to the route later followed by the railway from Port Pirie to Broken Hill.

Sam Pearce was keen on prospecting long before there was a Broken Hill. Some of the time he should have given to fencing and ploughing he spent in sinking shallow shafts in search of minerals. Six feet tall when such a height was unusual, and strong in physique, he worked hard at his prospecting, but found nothing. Nor was he a successful farmer. With little children to feed, he could not afford to cling to the farm in the hope that a bumper harvest would some day restore his solvency.

Early in the depressed year of 1893, Sam Pearce was

scratching for gold near Dashwood's Gully. There he met Will Brookman. The two men, failures in their first careers, and down on their luck in their more recent ventures, had little hesitation in accepting from George Brookman the offer of free food, a wage of £1 a week, and a share in whatever gold they might find. They were among the older prospectors who prepared to go west — Sam Pearce was 45 and Will Brookman was 33.

In June 1893 they sailed from Port Adelaide in an English mail steamer that was to call at the small port of Albany for coal. From Albany they must have caught the train to Perth, where they had the good sense to buy a copy of the mining regulations. Most prospectors relied on hearsay to tell them the exact rules governing how much mineralised ground each miner could claim. Brookman, who was really the travelling business-manager of the syndicate, had made his first shrewd decision.

At the inland town of York, Brookman and Pearce bought two horses, a spring dray, supplies and equipment. Setting out for the Coolgardie goldfield, they walked beside the light dray — it had no space for passengers. On their fourteenth day on the track they reached Coolgardie. There, on about 29 June, they heard of the discovery which Paddy Hannan had divulged a fortnight ago. In the winter sunshine they set out again, on the track to Kalgoorlie. It was easy to follow, with all the boot prints on the soft soil and the marks of iron wheels and the hooves of horses; and alongside the track were the embers of dead campfires and discarded food tins.

At Hannan's find, they watched several hundred men working in clouds of dust, digging narrow holes and trenches, or dry-blowing the soil. Pearce and Brookman had one advantage. Possessing money and equipment, they did not need to search for gold in order to pay their way. Instead, they could concentrate on finding the kind of lode or reef that George Brookman, back in Adelaide, might float on the stock exchange. They learned what they could at Hannan's rush and then moved south.

Passing scattered trees they reached red-brown hills three or four miles to the south. Here in the hard-rock country the two men had another advantage. They possessed something that the men who travelled lightly did not own: a dolly pot made of iron. Worked laboriously by hand, the dolly could grind down a small piece of rock, enabling the resultant grains of sand to be tested to see whether they included a few grains of gold. To his delight Pearce found that some of the unlikely-looking rock carried gold. One reef, mainly of quartz rock, was promising enough to prompt Pearce to use pick and crowbar to probe several feet into the reef. Again the dollypot was put to work. Though the gold in the hard reef was not visible to the naked eye, it was readily disclosed by the process of crushing.

Brookman set off for Coolgardie to register his claim to an area of ground measuring 300 by 300 yards — roughly the size of three large cricket grounds. He formally registered it as a protection area, thus securing the right to drive away any trespassers. While in Coolgardie, he wrote a short letter to his brother in distant Adelaide, reporting that they had found gold in 'one of the most mineralised hills in the vicinity of the new rush'. To enable Pearce to test the rock, Brookman paid for dynamite and hand-drills in Coolgardie and carried them back to their ground.

A prospector who found gold was usually over-enthusiastic. Will Brookman's letter, however, did not bubble with enthusiasm. It was probably read with caution by his brother in Adelaide. While specimens of stone carried the impressive assay of 100 ounces of gold to the ton, such assays were common on a new field. A prospector usually tested the prettiest specimen — it was bound to be rich. Such gold was like a needle in a haystack, and not easily found in any quantity when the mine was eventually opened.

Pearce and Brookman packed a box or case with specimens of the best rock they had found, firmly secured it, and sent it away by horse cart. Some weeks later, in Adelaide, the leaders of the syndicate opened the box at their office in Australasia Chambers, on the west side of King William Street. The carefully chosen specimens did not appear attractive. Where was the gold? The rocks were taken with some anxiety to a trusted mineralogist in Adelaide, a Mr Parkinson. He must have assayed them, finding gold where none was visible. He was very pleased indeed.

More money had to be raised to enable the two prospectors to continue their work. George Brookman converted the original syndicate into the Coolgardie Gold Mining and Prospecting Co., W.A., Limited, and divided it into 1000 shares each with a nominal value of five pounds. Presumably most of the shares were issued free to the original subscribers, but several hundred were sold at 50 shillings each in order to raise working capital. The choice of the name Coolgardie was revealing. The mine was not at Coolgardie, but that name was better known than Hannan's. Maybe, too, the name of Hannan did not sound right in such a proudly Protestant city.

The Coolgardie company was as much a circle of friends and fellow-religionists as a business firm. Most of its shareholders lived in Adelaide and probably knew one another personally. R. M. Gibbs, recently investigating these shareholders, found that many belonged to the same religious sect — the influential Congregationalists. On Sundays many of the shareholders could be found in the pews of the suburban Congregational churches at Brougham Place and nearby Medindie. There too could be seen the two leaders of the syndicate — George Brookman and George Doolette. It was typical of this rather tightly-knit group that it entrusted the Brookmans' old father with the printing of the new share certificates. One of the first certificates went to a Congregationalist pastor, who had given his approval by privately taking up shares. By October he was counting his blessings. One share bought for 50 shillings was now worth twenty times as much.

Back on the infant goldfield, Sam Pearce was claiming more ground for the handful of shareholders in Adelaide. As he walked over the low hills he was excited by many of the rock formations. Fortunately he and Brookman, like Hannan before them, kept their secret. The camel trains busily carrying water from the salt lake in the south to Kalgoorlie in the north passed close to the camp of the Adelaide prospectors, but those walking this track saw nothing unusual. Clumps of trees partly concealed the trenches dug here and there. The prospectors thus had the chance to explore the ground before outsiders realised what they were doing.

Pearce and Brookman were now confident enough to begin selecting a few large areas of ground, each ranging between 15 and 30 acres. Will Brookman apparently supplied the name for the first of these infant mines. He christened it the Ivanhoe, borrowing the name from the title of Sir Walter Scott's famous novel. Ivanhoe was also the name of the big house soon to be built by brother George in the Adelaide suburb of Gilberton.

One afternoon Sam Pearce found another tantalising mineral formation that deserved a special name. He was strolling, he recalled, with eyes fixed on the ground: 'After reaching the top of the ironstone hill, I came on a slaty-looking green rock sticking up six inches out of the soil, and there was a glint of gold showing.' He took the biggest lump of rock to the camp and thought hard about the wisdom of pegging the surrounding ground because every lease called for an annual payment of rent to the government. Next morning, his mind made up, he hammered wooden pegs into the ground and pinned a notice announcing that he was applying for 24 acres of mineral lease.

Brookman christened this latest mine after a little mine he had been working back in Dashwood's Gully in South Australia: the Boulder. Here at last, in far-off Western Australia, was a mine worthy of that imaginative name. He called it the Great Boulder. After the two men had used explosives to blast a shallow trench across the outcrop of the Great Boulder, they sent a confidential telegram to Adelaide to report that this immense ironstone hill carried rich reefs 'impregnated with gold'. The next lease pegged by Pearce was called the Lake View. The name was more pragmatic: the hill overlooked a salt lake.

Already Pearce had selected about half of what proved to be the richest ground on the field. Few people, however, took much notice of him. Diggers used to visit each other for a yarn, and those who came through the scattered gum trees to the camp of Pearce and Brookman were not impressed by the specimens of gold-bearing stone shown to them. The gold was hard to detect with the naked eye. Even the supposedly rich specimens actually resembled road metal. They were humdrum compared to the gold-spangled samples found at the north end of the same field.

The Great Boulder and other southern mines were

THE GOLDEN MILE

By 1894 the goldfields around Coolgardie were speckled with prospectors' camps, hidden in the bush. Rough sheds roofed with the leaves and boughs of trees were the normal protection from the intense heat. Armchairs or 'deckchairs' were made on the spot, using rough sticks of wood with hessian for the seat.

examined by W.H. Corbould who had studied in the Ballarat School of Mines and was familiar with Broken Hill and several other fields. Despite his youth, W.H. Corbould probably had the best formal mining qualifications to be found in early Kalgoorlie. The agent for South Australian investors keen to buy untested mines, he saw the Ivanhoe and the Great Boulder and frankly rejected them as not worth touching. He dismissed loquacious Pearce and silent Brookman as new chums. 'Both seemed to me extremely inexperienced in relation to geology and mining,' he said later. His view was the common view. Fortunately he learned his lesson and 30 years later was one of the first to appreciate a new Kalgoorlie, the silver-lead deposits of Mount Isa, of which he gained control.

At Kalgoorlie, Sam Pearce was commonly said to be an amateur prospector, but the verdict was unwise. Discovering gold in unexpected host rocks, Sam kept an open mind. There could be no higher virtue than an open mind in early Kalgoorlie, where honoured theories were easily turned upside down. As Sam found gold in unexpected places he quietly kept on exploring, often in rocks that were dismissed by others. A chatterbox, his prattle about the reefs and his theories of how they originated helped to turn people away from a calm evaluation of his gold leases.

Later it was admiringly said of Pearce that he could 'smell gold', but the tribute was too generous. Kalgoorlie was his only success in a lifetime of prospecting. Part of that success came from the fact that, by the standards of the time, he was a novice. Nearly all the experienced men on Kalgoorlie had come from the eastern colonies where the trusted axiom was that 'quartz was the mother of gold'. So they were less impressed with the southern end of the field where Pearce worked than with Hannan's northern end where quartz rock was plentiful. Pearce simply trusted his eyes. It did not worry him in the least that he was finding gold where in theory it should not exist. In his behaviour he echoed the old Cornish saying that 'gold is where you find it', not where theory insists it should be. He became well known for his faith in the ironstone — a rock that was rarely associated with gold in the chain of goldfields along the Great Dividing Range of eastern Australia. Often he said, 'the iron cap covers the golden head'.

Inch by inch the knowledge was pieced together, painfully and at high expense. The main goldfields of Western Australia were in Pre-Cambrian rocks, which were much older than the rocks in the eastern half of the continent. Most of Western Australia's gold was to come from the greenstone belts that cover only one twentieth of the area of the state. In Kalgoorlie most of the gold was to be found in a greenstone formation called Golden Mile Dolerite. Within the dolerite the gold was not evenly spread, but rather concentrated in many orebodies, some very large and some very small. Few of these orebodies poked their heads above the surface of the ground. Sam Pearce and Will Brookman happened to find a few of them, of great magnitude, that lay on or near the surface. To their credit, the pair were pragmatic and shunned the theory which insisted they were wasting their time.

While the bulk of Kalgoorlie's gold lay embedded in dolerite, gold was also found in other kinds of rocks. Rich gold was to be discovered in bands of slate or in porphyritic rock. Very rich gold was to be found in basalt — that is, in volcanic rock.

To the experienced geologist it was a perplexing place, and to some degree it remained so. In 1904 one of Australia's finest mining reporters, Donald Clark, expressed in his long book *Australian Mining and Metallurgy* his puzzlement when he descended the 1000-feet-deep shaft of the Kalgurli Gold Mines. He was astonished to see everywhere 'a bluish-grey rock'. To his eyes it looked exactly the same when it contained gold as when it contained none. He reported with surprise that between the 640-feet and 920-feet levels (a big vertical distance in the mines of that day) 'no free gold is visible', and yet the gold was there, invisible.

In the deeper or sulphide zone of a typical Kalgoorlie mine, unlike those in Victoria, the experienced eye was unable to tell with sufficient accuracy whether the ore was rich, moderate or unpayable. And so in the course of a year, many thousands of chemical assays of specimens of rock from the mine had to be made to make sure that big blocks of ground were worth mining.

Only the northern end of the goldfield seemed familiar. There, not far from the site where Paddy Hannan first pitched his camp, most of the gold was found in the strong quartz veins of Mount Charlotte. More like a Victorian gold mine, it delighted many miners fresh from Ballarat, Stawell and Walhalla. It the following eighty years, however, barely one per cent of Kalgoorlie's gold would come from that northern end of the field. Today, at last, that northern end is fully appreciated.

In retrospect there were two major discoveries on Kalgoorlie. Hannan made one and Pearce the other and more important discovery. Sam Pearce's achievement was to perceive that the quartz-dolerite-greenstone was friendly. Probably this was more important than anything Hannan himself discovered. One day a statue should be erected in Pearce's honour at the south end of the field. In one hand should be a waterbag, in the other a dollypot, and his mouth should be open, for often he talked to himself, convincing himself that the greenish slate-like rock was rich.

Just before the hot Christmas of 1893, those who visited Pearce's end of the field found a simple rural existence. Tents — grimy with red dust — stood among the trees. Bough sheds offered shelter from the hot sun. A few wheeltracks twisted their way between the trees. Every month the trees became fewer as they were chopped down to supply firewood for cooking, and bush timber to line the shafts or build the miners' huts.

In places where the natural bush persisted, the scenery

had a charm that surprised those newcomers who imagined that they were entering a desert. Kalgoorlie, and a small region around Norseman and Kambalda, were the only places on earth to grow the flowering eucalyptus called Coral Gum. Its dark grey bark and lacklustre leaves gave no sign of its majesty in blossom time. Then the tree blazed into delicate or vivid pink.

The trees were more capable than human beings of surviving in this arid climate. Water was unbelievably dear in the first summer. Even a cup of tea was a luxury. Miners did not wash their clothes in water: they beat them with a stick to remove the red dust. How the plates and knives were cleaned was a question that sensible visitors did not ask when eating their first meal at the camp of a neighbour.

Small bush flies were a pest, and the ants made a black trail to any food exposed in a hut. Fortunately there were no mosquitoes. As summer arrived, the heat bewildered those not accustomed to it. For seven days in a row the temperature in the shade could pass 100 degrees Fahrenheit. The nights were cooler. The night sky was clear and the stars were brilliant when there was no moon and the miners who sat in front of the cooking fire after the evening meal could feel a deep contentment. A month before the first Christmas, a letter in the Adelaide *Observer* describing daily life near the camp of Brookman and Pearce caught this mixture of excitement and lethargy. A six-foot snake had been killed near one of the tents. After dark an old mopoke could be heard singing on the branch of a dead tree. The men slept on the bare earth, on a pile of gum leaves which gave off 'a beautiful aromatic odour'. At about 5.30 in the morning, each man took his turn to light the fire and cook the breakfast of porridge, damper and jam, maybe bacon or tinned meat, and black tea. It was wise to eat before the flies became busy, and to start work before the sun gained strength.

No steam engine was yet working at a mine. A makeshift blacksmith's forge with hand bellows was the most advanced technology to be seen. No shop, no grog shanty, no public amenity existed in this part of the field. No woman's voice was heard.

At the mines found by Sam Pearce, little gold was won during the first year. Pearce dollied some of the very rich specimens of rock and sold the gold to help pay the rent on the mining leases. More money was needed to pay for the mounting expenses. The head of the syndicate, George Brookman, tried to raise it in Melbourne, the financial heart of Australia. Floating his first mine as a separate public company, the Ivanhoe Gold Mining Company No Liability, he issued 30 000 shares. Half were offered to new shareholders for cash, and the other half went as a reward to the shareholders in Brookman's Coolgardie Gold Mining and Prospecting Co. In a prospectus dated 25 September 1893, he quaintly announced the company's place of intended operations as 'Hannan's Rush, Coolgardie, Western Australia'. Hardly anybody in Melbourne had heard of Hannan's.

This was one of the finest investment opportunities in the history of gold shares in Victoria. Here was a mine valued by Brookman at just over £10 000 — an absurd price, in retrospect, because the company would soon pay more than £3 million in dividends. Victoria's investors did not rush the offer to enter Kalgoorlie on the ground floor. Likewise they shunned the northern end of that field when, just before Christmas, the Maritana Gold Mining Company was launched to mine gold at a place mis-spelt as 'Hannen's Rush'. Before the year was over Brookman had raised further small sums by floating two other leases into small separate companies — Lake View and Great Boulder. Kalgoorlie was still hopelessly undervalued by the public at large. At the start of the 1894 the market valuation of all the gold mines at Kalgoorlie was far lower than the valuation placed on each of at least forty different goldfields in Australia. On the stock exchanges Kalgoorlie was a nobody compared to Bendigo and Mt Morgan.

George Brookman had such trouble in raising capital to develop his syndicate's mines that he thought of trying overseas. Gold mines, especially those of the Transvaal, were booming on the London exchange. If South African mines, some of little worth, were raising big sums in London, surely West Australian mines of high promise could raise capital there. In March 1894 his fellow-shareholder George Doolette, the former owner of an Adelaide outfitting and tailoring house, reached London with plans to gather more capital for equipping the Great Boulder mine. In his mission he was helped by Zebina Lane, a burly confident engineer who was consultant to the busy middlemen arranging the sale of Kalgoorlie mines to new London companies. Lane had been brought up on Victorian goldfields where his Canadian-born father was successively miner, blacksmith and mine manager. The son Zebina once claimed that at the age of 15 he himself first became a mine manager — but he was known to make the occasional loose statement, even on company prospectuses. An early arrival at the Broken Hill silverfield, he became mayor of the town before moving west, where he was probably one of the first mining experts to see magic in the Great Boulder.

With the help of Zebina Lane, the Great Boulder Proprietary Gold Mines was floated in London in June 1894. The public, not fully convinced, subscribed only £30 000 of the £40 000 sought. Zebina Lane, with an airy wave of his hand, said the ore would average five ounces to the ton. It was enough to make people suspicious. Even so, Great Boulder now had the money to buy machinery and so become a gold producer.

A long line of promoters began to launch Kalgoorlie mines on the London stock exchange, raising big sums for themselves as well as money for exploring and equipping their mines. The Pearce discoveries that had begun as little Australian companies were refloated as big London companies, providing in the process more profits for the original Australian investors. Zebina Lane, fresh from his

IS THAT GOLD?

'On the stock exchanges Kalgoorlie was a nobody' but it had the nucleus of its own stock exchange even while tree stumps still stood in the main street. J. Howard Taylor erected an iron shed and put up his sign 'SHAREBROKER'. Aborigines, visiting and local, were assembled for the photograph.

success with Great Boulder, helped to promote other Kalgoorlie mines, including Associated Gold Mines of W.A. and Great Boulder Perseverance. Several of the mines he helped to float were wonder mines, but later he developed a kind of promoter's lockjaw, nodding approvingly at the mere mention of the name of dubious Western Australian mines. He died in Berlin in 1912, a rich man.

Nearly every gold-mining lease that lay within gunshot sound of Great Boulder was bought from the original owners at a high price and sold in London to a new company at a far higher price. Thus a few miners from the South Australian copper town of Moonta held a lease called the Golden Pebble. Whether it then held even one pebble of gold was open to dispute. Late in 1895 they cleared the large sum of £8000 by selling their mine to a London promoter. Worthless mines raised a lot of money in London. Madness was in the air.

Most London investors treated Kalgoorlie as a gambling counter. They bought shares in Great Boulder or Iron Duke in the hope that shares would rise and they could then sell them at an early profit. Adelaide supplied most of the gambling counters. Many of its citizens held thick bundles of shares in the big Kalgoorlie mines. As late as 1896, 70 000 of the 175 000 Great Boulder shares were listed on the Adelaide register. Increasingly they were sold to English speculators. If a balance of payments could be calculated for South Australia in the years 1894 to 1896, one of the biggest items would have been the income earned from the sale of Kalgoorlie shares.

George Doolette, watching the share market from his new home in England, vowed that it was like a story from 'the Arabian Nights'. Having bought a large property at Caterham in Surrey, he entertained lavishly on his lawns and by his lake and rustic bridges. At a garden party in June 1896 the rising young Gippsland contralto, Ada Crossley — 'another Clara Butt', critics were to say — was to sing to Doolette's guests. It was appropriate that she should entertain that charmed circle of Kalgoorlie shareholders, because she was one of them.

Sam Pearce, the finder of these deposits which kept the London printers of coloured share-certificates so busy, was another of the fortune-makers. He deserved his small fortune. He had been instrumental in pegging a more valuable block of minerals than any previous prospector in Australia. He had done his searching, theorising, pegging and the initial exploring with calm competence. But his health was now suffering from the hard life and poor food of the primitive goldfield. Moreover, mining captains were replacing prospectors, and Pearce was essentially a surface prospector. The main lode at Great Boulder was actually found after his departure.

Pearce returned to his large family in Adelaide and lifted them above the daily struggle. After buying a grand house, he set out on a world tour with those of his family who wished to travel. But he remained a prospector at heart. He must always be searching. The Klondyke gold rush in

'IS THAT GOLD?'

Ada Crossley, a star of European opera, visited the Golden Horseshoe mine in 1904. She posed dramatically, as well she might, being the possessor of shares in the Golden Mile as well as one of the world's finest contralto voices. Next year in London she married the son of one of the Adelaide families enriched by Great Boulder. The narrow-gauge tramline can be seen curving into the background. Along those rails the truckers pushed the small iron trucks loaded with ore, and then the cage or skip carried them up the shaft to the land of daylight.

the cold north of North America attracted him as it attracted scores of Kalgoorlie men. He searched in Mexico and Zululand — he was always tempted by new mineral fields. Even at the age of 80 he was still tempted. A forlorn mining field near Adelaide caught his fancy, and for a time he lived in a tent, cooking his stew or porridge in an iron pot, and stirring it with a 'claret ladle' made of sterling silver. He died in the Adelaide hospital in 1932, and in the following week much was written about mining's good luck and bad luck, in both of which he amply shared.

When Pearce left Kalgoorlie it was not yet producing much gold. Rather, it was making fortunes — so long as the share market was rising — for many of those who bought and sold shares in London. In England's craze for the latest information about each mine, the vital newsline was Kalgoorlie's makeshift post office, which sat in a corner of the general store owned by Faiz and Tagh Mahomet. The telegraph line arrived in the winter of 1894, and it instantly became the hearing aid of the London stock exchange. After a rich lode was found those with knowledge could make a fortune by buying shares in London, before the news of the discovery arrived. On some days the little telegraph office with the iron roof was jammed with people struggling to gain priority for sending telegrams to companies, individual sharebrokers and investors in Adelaide and London. The outwards line could take only one telegram at a time, each word being tapped out by the operator. In those weeks when London's share market was feverish, 20 hours might pass before the telegram was sent, and another two days could elapse before it reached London.

The long-term value of Kalgoorlie's mines was far from certain. The early share-buyers had no means of knowing more than say 1 per cent of what is now known. Knowledge of the depth and extent of the gold at the rich southern end of Kalgoorlie came slowly. Nobody had more than a semblance of a theory which could explain where, within each lease, payable ore was likely to be found on this puzzling goldfield. Nobody yet could possibly know — though several claimed they did — that here was the continent's biggest goldfield.

The worth of Kalgoorlie's companies could not be fully evaluated until they were actually crushing the rock and extracting the gold. Traditional crushing machines — batteries or stamp mills — were ordered by the Ivanhoe and Great Boulder and several other companies, but the difficulty was to cart the heavy machinery inland to the mines. In the summer of 1894–95, horse teams could be seen struggling with heavy loads along the inland road. On the dry stretches a pile of chaff and hay was waiting for the teams — almost the most expensive bags of chaff in Australia, for they had been carted long distances in readiness and placed under guard. At last at the south end of Kalgoorlie, the horse-drawn wagons unloaded their iron boilers, engines, heavy cast-iron stampers, and the amalgamating tables. Occasionally a vital part was missing when the unloading was completed. To await a replacement from the foundry in Adelaide or Gawler was to waste months. At the mines, handymen came to the rescue. When one iron part was found to be missing, a makeshift block of wood was hacked from the stump of a gimlet tree and shaped into a temporary replacement.

The most important battery was erected by one of the Brookman companies, the Leviathan Quartz Crushing Company, which treated ores from many mines in small batches on the shores of Hannan's Lake, about seven and a half miles from Kalgoorlie. Those walking towards the Leviathan battery could not lose their way. The roar of the stampers was like a barking guide-dog. As visitors came closer to the noise, the thin pencil-like chimneys came into sight, and the high stacks of firewood with long untidy lengths protruding here and there. They saw, too, the glare of the salt heaps: salt was a waste product of the steam-driven plant which expensively turned salt water into fresh.

Everybody within a radius of several miles knew when the Great Boulder battery was being tested. The thump-thump of the iron stampers as they initially crushed samples of rock from the mine was distinctive. The sound was often followed by silence. Sometimes the silence lasted for a week as repairs were made. With an unfamiliar ore like Great Boulder's, it was not easy to crush the lumps of ore down to grains of roughly the right size, and to separate the few grains of gold from the innumerable grains of barren rock. Moreover, salty or muddy water used in the treatment plant caused trouble.

At Great Boulder, the new battery was about to begin continuous production in April 1895. Zebina Lane was out from London to supervise such an important event. A few hundred tons of rich ore had been placed in jute bags and sewn up with thread — and even clamped with a seal — as a security against the theft of the richer specimens. To satisfy the stock market a rather rich crushing was deemed necessary. Initially a total of 128 tons of ore, most of which seemed poor in the eyes of a reporter from Coolgardie's *Goldfields Courier*, was crushed in the battery. To the surprise of some observers it yielded 1289 ounces of gold, or ten ounces to the ton. Great Boulder had passed its first test. By the end of the year its tally of gold was worth £107 000.

Kalgoorlie was not yet accepted as the main goldfield of Western Australia. No single field could yet claim the crown. The government geologist, Harry P. Woodward, exuberantly claimed in the middle of 1895 that the salt-lake district of Western Australia was the 'most extensive gold district known in the world', and so it was premature to say which field would be the greatest. Randolph Bedford, then a Melbourne mining journalist, asserted in his own coloured magazine *Clarion* in November 1896 that 'Westralia' — he helped to popularise this catchy name — was the world's biggest mining field. He went on to predict that 'Westralia will be dotted with cities living on

The day of the official opening of the Leviathan battery drew a crowd of investors and mining engineers to the remote site. The day must have been cool — many of the men are wearing waistcoats or pullovers. Will Brookman, the prospector, is one of those in the top right corner.

IS THAT GOLD?

Bare and dusty Hannan Street, Kalgoorlie's main street, was becoming busy in 1895. The scene was only five minutes' walk from the spot where Patrick Hannan, two years earlier, found the first gold.

reefing when generations hence have become impalpable ashes'. He could not conceive that Kalgoorlie would be the only reefing city worth of the name.

In the months when Great Boulder was at last showing its riches, many investors choosing the best goldfield still singled out Coolgardie. It was older. More was known about it. Its deepest shaft was deeper and was said to be still showing gold. In the winter of 1895 Coolgardie's total output of gold was far ahead of its new rival. Within half a day's buggy-ride from Coolgardie, mines such as the Londonderry and the Wealth of Nations yielded specimens of gold-studded rock that took men's breath away. In that era part of the attraction of native gold was its sheer beauty, and Coolgardie was now yielding eye-catching specimens that delighted connoisseurs. By contrast, Kalgoorlie's richest ore often looked like road metal. Harry Woodward, in his guidebook for gold seekers, was still guarded about Kalgoorlie and gave the lukewarm view in June 1895 that 'it will not be long before very good practical results will be obtained'. If he had personally chosen shares for himself, he probably would have favoured the Coolgardie district and the Londonderry mine, 'which everyone allows is one of the most promising lodes ever opened'. Sadly, a year later, it was not as promising as a dozen different Kalgoorlie mines. Coolgardie as a gold district jilted thousands of its lovers.

In the spring of 1895 a journalist and artist, Julius Price, arrived in Kalgoorlie on a brief visit. With his quizzical look, curly hair and waxed moustache almost spiked at the ends, he was viewed with curiosity but eagerly welcomed, especially when it was heard that he would describe the new goldfield for one of London's leading journals. Every now and then he stopped to sketch in black and white a string of camels, or trees growing pell-mell in the middle of Kalgoorlie's main street, or a horse turning the whim at the Brownhill mine. These sketches appeared in the *Illustrated London News*, along with his regret that he was awakened in the night by the 'hoarse screech of some steam-whistle calling in the night shifts'.

To the delight of the tens of thousands of shareholders in Kalgoorlie's mines, Price had something important to say about the mines themselves. He informed a large London audience of what several other writers, rightly or wrongly, were beginning to assert: that this dust-coated town possessed the two best gold mines in the world.

CHAPTER THREE — UNLOCKING THE JEWEL CHAMBERS

THOSE WHO RAN the post office at Kalgoorlie knew how quickly the population was increasing. After the mail coach arrived and the letters and posted newspapers were sorted, several hundred people jostled inside and outside the iron building, asking for their mail. Some came not only to collect letters but to remit part of their wages back to their families. Nearly all the workers on the goldfield came in steamships from Victoria and South Australia in the hope of escaping the depression. Kalgoorlie was settled from the ocean.

In Ballarat day after day in the summer of 1895–96 the newspapers reported that people of the district were preparing to go to the faraway west. Correspondents from the nearby towns sent to Ballarat small items of news describing the departure of their townsfolk and the farewells. From Ross Creek a band of residents was about to go west. Sebastopol, reported the *Ballarat Courier* in January 1896, would soon see a 'great exodus of miners'. At Clunes, once the most bustling quartz town in Australia, 30 people 'if not more' would leave this week, joining the hundreds already in the west. At the little gold towns of Egerton and Gordon, staid residents were leaving along with the single men and young married men. In Ballarat, a militiaman was farewelled with the gift of a gold locket before he went west in the steps of other militiamen. Bid farewell at the railway stations by people they might never see again, the largest group of these gold seekers or job seekers was bound for Kalgoorlie.

Coastal steamships, slick or grubby, crossed the Great Australian Bight to Albany or Fremantle in a procession marked by black smoke. In January 1896, in the course of a dozen days, the Adelaide Steamship Company alone sent four ships away from Melbourne, their deck-rails lined by gold seekers. Such shipping firms as McIlwraith, McEacharn & Co., Huddart Parker, Howard Smith, and other Australian shipping lines concentrated most of their ships on the route to Western Australia, conveying up to 600 passengers in each packed vessel. Every week at least one larger overseas liner about to sail from Melbourne offered more expensive berths to gold seekers bound for Albany.

Forty years later the oldtimers at Kalgoorlie, reminiscing about early days, would talk eagerly of the voyage across the Bight, the first day of seasickness, the armpit-odour of the airless sleeping quarters where as many as four bunks were fastened one on top of the other in order to hold as many as 200 passengers in a single dormitory. They recalled, when the night was balmy, the concerts and community-singing on deck, with the funnel smoke swirling past the seated passengers. As long as they lived they remembered the name of the ship in which they set out to cross the Bight.

The journey of some 2500 miles from Ballarat or Bendigo to Kalgoorlie was not planned lightly. Money for the adventure had to be saved. The cheapest fare from Melbourne to Albany usually cost a man the equivalent of a fortnight's pay. The train ticket cost perhaps another week's

THE GOLDEN MILE

Newcomers from the east proudly announced their background on their Kalgoorlie shops and offices. The Semaphore chambers were erected in about 1895 by a young man who came from the Adelaide suburb of that name. The owner, standing fourth from the left and right behind the cart wheel, was W. H. C. Lovely, a mine manager and assayer.

wages. Then at the temporary terminus of the railway line, out in the Never Never, the passenger alighted and counted his money and made another decision. If he boarded the horse-drawn coach he handed over the best part of another week's wages. If he decided to walk the remaining part of the journey to Kalgoorlie he found himself paying a few days' wages to the owner of the horse wagon who agreed to carry the swag weighing, say, a hundredweight.

The first train from Perth finally reached Kalgoorlie on 8 September 1896, a day when mine-whistles blew again and again in celebration. The journey by ship and train all the way from the far side of the continent was now quicker but still dear. A typical Victorian had to save the equivalent of one month's wages before setting out for Kalgoorlie, and that left nothing for tobacco, beer, and meals on the train. It was no wonder that in the pioneering years few of the married men brought their wives and children.

In Kalgoorlie the newcomers found a job as soon as possible. If they had no experience of mines but were strong they began work below as truckers, or worked as labourers in the mills. For years to come they were willing to accept almost anything as a residence — a shed of iron, a tent, a hessian lean-to, or a room in a boarding house. My grandfather, Sam Blainey, arriving from Eaglehawk in Victoria as an engine driver, camped near Kallaroo railway station in a tent augmented by wooden boards ripped from a large packing case in which one of the first diamond drills had arrived.

The township was an eyesore. The main streets had no footpaths and no gutters. Corrugated iron, unpainted, was the chief building material. Until the railway arrived from the coast the iron was so dear that a veranda in front of a house or a shop was a luxury. Hundreds of men lived in tents or occasionally a shelter of leafy boughs. The town looked desolate, and the few women who arrived must have wept or tried hard to suppress their tears. To look down on Kalgoorlie from a nearby hill in 1896 was to see almost every building in the town, for there was hardly a tree, shrub or veranda to conceal them. When seen from a distance on the dusty plain these little boxes, with no veranda and no fence, resembled a large flock of straggling sheep. Randolph Bedford, writing his mining-and-travel magazine the *Clarion* in February 1897, was harsh towards two of the goldfield towns: 'Kalgoorlie and Menzies take the cake for dust and general insanitaryness.' He did not exempt Perth and its hotels and restaurants. He called them 'unspeakable'.

The goldfields had one big attraction over Melbourne and Sydney. It was possible to save money in Kalgoorlie. The wages were perhaps double those paid for a labouring job in Melbourne, and so half could be saved and sent home through the post office. In an era before unemployment benefits and old-age pensions, the money orders sent from Kalgoorlie provided the bread, meat and boots for thousands of families in fading gold towns in Victoria.

One acute disadvantage was the continuing expense of

THE GOLDEN MILE

After the steam train came a few steam-driven traction engines that travelled slowly on the roads around Kalgoorlie. This engine was hauling the heavy iron stampers of a battery or crushing plant from Kalgoorlie to a remote gold mine at the end of this sodden track. This mill had only ten stampers, but the Golden Mile had treatment plants where as many as 150 stampers made a deafening noise.

water for washing and drinking. On most mines the men received not only a daily wage but a daily allowance of water. Two gallons was the ration at the start of the shift and two gallons to take home at the end of the shift. The water was usually condensed from salt water at high cost, the salt water coming from lakes nearby or from shafts sunk on the mining lease in order to tap the water found in some abundance at a depth of about 200 feet. Many of the mines sank a special shaft solely to reach this rainwater which had been percolating down from the surface in past centuries. Many companies owned shafts in which the water was 20 to 25 feet deep. The shaft in Hannan's Golden Pebbles found no gold but something equally as valuable: 25 000 gallons of water were sold each day.

By 1897 mains carrying a precious supply of expensive water ran along Kalgoorlie's busier streets. The desalted water, produced in a water factory topped by three black chimneys, was piped to shops and the wealthier houses by a private company. Nearly every drop of dirty water was saved for the little trees and geraniums living in the few brave gardens. In the hotels the bathroom was locked. Guests who were granted permission to bathe noticed that the water barely covered the bottom of the metal bath. They must have blinked when they saw the bath water separately itemised on the hotel bill. The cost of a bath was outrageously high. It paid to remain dirty, though that idea was less popular when more women settled on the goldfields.

The lack of cheap water impeded the fighting of fires. In Kalgoorlie on the evening of 9 October 1895, those people who were out of doors saw a bright glow to the west. Later they learned that the powerful glow was a fire raging in busy Coolgardie. A lamp had been knocked over, a property had been set alight, and buildings covering a whole acre block were burning in one of the most damaging blazes in Australia that year.

In the wide main streets of Coolgardie new buildings quickly filled the charred gaps. The town had the look of an infant capital city before its civic leaders began to understand that it had no sure future. Coolgardie, so proud of its miraculous progress, was realising that it had one grave deficiency for a gold town. It lacked gold at depth.

In the year 1896, Coolgardie's gold output was passed by Kalgoorlie. A year later Kalgoorlie was far ahead. Curiously, the glory remained with the older town. The official name the government in Perth gave to the Kalgoorlie goldfield was the rather belittling name of East Coolgardie. This was to remain its official name, long after it became Australia's greatest goldfield. It was almost like giving Canberra the formal name of Sydney South.

Victorian critics were not yet impressed with Kalgoorlie. They said it was an upstart, suffering from delusions of grandeur. It was rich, but would it last? Here perhaps was another Coolgardie, rich on the surface and poor at depth? In the autumn of 1896 the deeper shafts at Kalgoorlie

Some of the working miners who helped to answer the question, will the gold live down? They gathered for this photograph in 1895 at the main shaft of the Golden Horseshoe mine. The miner sitting on the far left carries his candle — the main source of light in the early mines. (Photo from Battye Library, no. 10221 P)

were not much deeper than 200 feet. In contrast, Bendigo was mining gold in shafts 15 times as deep. Kalgoorlie, said critical Victorians, was largely an untried field. Its monthly output of gold was not yet as large as that from Bendigo. Its biggest mine was no larger than many Victorian mines. Thus each month for some 17 years the Long Tunnel mine in Walhalla, Victoria, had been turning out as much gold as the best Kalgoorlie mine was producing. Only in April 1896 did Great Boulder produce more than 4000 ounces for the month, thus surpassing the Long Tunnel — the source of some of its best miners.

In Kalgoorlie, as in numerous new fields, the enthusiasm and the boasting skipped far ahead of reality. Victorian gold towns, fearful of the loss of so many of their best men to Kalgoorlie, thought it was time to fight back. Who was this western upstart? The Melbourne *Age*, with the biggest readership of any daily newspaper in Australia, sent an experienced mining journalist to the distant goldfields. On 17 July 1896, in the first of a series of articles, he began with a declaration of war: 'It is quite time that the public should know the real unvarnished truth about Western Australia.' He thought it his duty to remove the golden varnish from Kalgoorlie. Its mines in his opinion did not 'present conditions associated with permanency'. He was not dogmatic, just sceptical.

He pricked the inflated opinions held about Kalgoorlie's glamorous mines. To those who boasted that the Ivanhoe had the richest reef so far found in Australia, he replied that they 'do not know what they are talking about'. The London share market, he lamented, was delirious. To value the whole Ivanhoe mine at £460 000 was sheer exaggeration. On the evidence of his own eyes the Ivanhoe mine could never produce that profit. In his view the Lake View Consols mine (soon to become a sensation) did not deserve the praise poured on it in London. But at least he was correct in implying that most of the hundreds of small companies and syndicates, busy or idle, near big Kalgoorlie mines had not found enough gold to fill a small cavity in a tooth. Their shares were sometimes riding high simply because they were near or said to be near a mine actually producing gold.

A map of the mining properties at Kalgoorlie was like a map of tiny subdivided hobby farms. These mining properties were mostly square but many were oblong and some were crooked. A few were six-sided and one was even nine-sided. Their very names suggested how much their appeal depended on proximity to gold-producing mines rather than on any merit of their own. As late as 1900, by which time many of the small companies had expired, the Great Boulder mine was ringed by a halo of lesser imitators. Each of these lesser companies had raised money on some distant stock exchange by claiming it was next door to the rich lode. Thus at the northerly end of Great Boulder stood such 'mines' as North Boulder, Boulder North Extended, Boulder Central, Boulder Central Extended, Central and West Boulder, and Brookman Bros

Boulder Co. It was confusing to the newcomer who wished to find his mate from Ballarat or Broken Hill but carried in his head only the knowledge that the mate was working in a mine called the Boulder. When he asked for the way to the mine, people said, 'Which Boulder?'. If the newcomer decided to walk beyond the Great Boulder he found another cluster of mines called Boulder. Walking further south he learned that there were mines called Great Boulder South, Boulder Half-Mile South, the Boulder Bonanza and at least five other mines with some version of Boulder written on the side of their office — if they had an office — or if they even had a mine!

The honoured name of Hannan also sprinkled the map. There were the well-known companies such as Hannan's Star and Hannan's Brownhill, but at least seven more mines also carried the name of the famous Paddy. These included Hannan's Golden Trees and Hannan's Golden Morning Star — a star which quickly sank.

The gold-bearing ground was now proved to exist, with here and there a gap, for a total of six miles on a north-south line. Optimistically, additional ground had been pegged out by eager companies for a total of ten miles. On an east-west line the optimism was just as fierce. Con Lynch, manager of the Ivanhoe mine and earlier a discoverer of gold on the west coast of Tasmania, called these outlying mines the 'salt bush mines'. In 1896 he would say, 'Cast your eye over there', and point his finger some two miles away to the little poppet heads of companies that claimed they were on the main gold-bearing lode: 'It is marvellous to think how the Boulder, Lake View and Ivanhoe reefs would have to twist about to pass through their leases.' These mines, in Lynch's view, were so far into the saltbush country that they were worthless. Even 90 years later, most of those leases have produced not one ounce of gold.

Speculation in mining shares was almost an epidemic in Kalgoorlie in these exploratory years. After dark, crowds of working miners, shopkeepers and tradesmen would gather in the main street of Kalgoorlie to watch the evening auction of mining shares. Kalgoorlie, like Coolgardie, had its own stock exchange, several informal stock exchanges, and dozens of sharebrokers. On some exchanges the sale of mining shares was conducted by what was called a Dutch auction, beginning at a high price and moving down in steps. At times the prices of the gold-mining shares were called in the open air, in front of the brokers' offices. The dry climate smiled on the selling of shares under the open sky. The young owner of Deland's Bakery explained in a letter to his parents in South Australia in September 1895 that whenever the mining exchange was held outside in Hannan Street on Saturday night, at least 1000 men were standing there, listening to the bidding. He himself was a speculator: he heard from his bakehouse the prices being called aloud and listened closely for the names of shares in which he was interested.

Kalgoorlie's investors studied the form-guides not only

'Speculation in mining shares was almost an epidemic in Kalgoorlie.' This open call or auction of mining shares is being conducted at the auction room of H. D. 'Larry' Pell, who at the age of 19 had been a stockbroker in Broken Hill before joining the rush to the west. The date on the photo is 1920. If true, it is very late for such a crowd of share-buyers. On the other hand it is clearly not a scene from around 1900 when most of the share-buyers would have been very young. The items suspended from the ceiling are kerosene heaters, for use in winter.

of local mines but of those hidden away in the surrounding countryside. In 1897 a little company calling itself the Bank of Ireland Gold Mining Co. was floated to explore a gold lease about twelve miles from Kalgoorlie. Its printed prospectus showed that the rage for mining shares was spreading to nearby towns. The sharebrokers listed as accepting applications for shares make strange reading today. While one came from Adelaide and one from Perth, three were from goldfields towns that have virtually vanished — Menzies, Kanowna and Bulong. In addition, one broker lived in Coolgardie and two in Kalgoorlie. There the share market was busy enough for some brokers to employ clerks.

Local investors had the advantage that, being on the spot, they were often first to hear the rumours or news of any discovery or any change in an orebody being mined. At the same time they were vulnerable to those Londoners who brazenly rigged the price for particular shares. For all investors, no matter where they lived, this goldfield remained largely a riddle. Its uniqueness combined with its richness made some experts think it would rock the world. The same uniqueness made others feel cautious.

In an assayer's shop in Kalgoorlie a clever discovery heightened the feeling that this goldfield was like almost no other in the world. Arthur Holroyd was a Melbourne-trained chemist who made his living by testing the rocks sent from smaller mines that could not afford to employ an assayer. In May 1896 a Kalgoorlie mine of scant consequence, the Block 45, sent him calcite rocks to examine. The darkish rocks did not appear promising, and yet when he tested them in a furnace they were rich in gold. Puzzled, he repeated the test. Again the result was a high assay of gold. The idea came to him that perhaps the presence of the rare element tellurium was camouflaging the richness of the rock. Placing a specimen of the ore in a bowl, he tested it with hot sulphuric acid and saw, as he had predicted, a carnation colour appear. His hunch was correct. He announced his findings in a letter to a Kalgoorlie newspaper. The discovery that telluride was present in many mines and was often rich in gold was so significant that a historical plaque was later placed at the site of the Block 45 mine, though the plaque has since been removed because of the encroachment of the big open cut.

In Australia, Kalgoorlie is the only field to be distinguished by the importance of tellurides. A few other Western Australian goldfields have small quantities; a few gold mines in Charters Towers, Gympie and Cracow in Queensland have it in traces. It was not surprising then that telluride, being unusual, was at first overlooked in the Kalgoorlie mines. In the underground workings, many miles of vertical and horizontal 'corridors' or streets had been created by the miners, and some — unknown to the mine manager — had actually cut through specimens of telluride. There it was, staring every passer-by in the face. Once the mineral was positively sought, it was found in at

UNLOCKING THE JEWEL CHAMBERS

Is it telluride? A scene at the 1600-ft level of the Great Boulder in 1903. The ore is about to be hoisted about one third of a mile up the shaft. This is one of the fine photographs collected by the Museum of the Goldfields at Kalgoorlie, and originally taken by local photographer Jack Dwyer, a native of the Gaffney's Creek goldfield in Victoria.

least 30 mines, big and small. It was even detected on the walls of shafts and drives that had been visible to the passers-by for several years. Now we know that about 15 per cent of the ore on the whole goldfield consisted of gold telluride. After Holroyd's announcement, the optimists were entitled to cheer. They could point out that, in this unique mining field, some of the richest ore had been thrown onto the dumps through sheer ignorance.

Alert miners now learned to detect the telltale colours of the different kinds of gold telluride. They especially noted the pale bronze-yellow of the calaverite, and the slippery, silvery look and the strong cleavage of the sylvanite and krennerite. Gold telluride mostly occurred in veinlets that crossed the lode at right angles, and quartz was often present too. Frequently the presence of telluride was a sign that the workings were entering a richer shoot of ore.

Rarely in the history of a mining field had a mineral, long neglected or even discarded, suddenly become the portent of richness. Remarkable tales of overlooked riches came to light. Thus miners in the Croesus, at the 170-feet level, found rock that seemed unattractive and low in gold. Specimens of the rock were sent to the assayer who reported that it carried an exhilarating 480 ounces of gold to the ton! Looked at again, the specimens were telluride-gold.

After the presence of this cinderella mineral was widely known, sharebrokers had another reason for booming the goldfield. The loudest boomer was a European mining expert, Modeste Maryanski. He was not all that modest. He claimed to have discovered the first tellurides, but he could not have discovered them because he did not reach Kalgoorlie until after Holroyd had made his announcement. We know when Maryanski first visited the Western Australian field because one of his fellow passengers between Melbourne and Albany was Henry Lawson, then rising to fame as a writer. Lawson wrote that the Polish minerals expert on board was 'a stout, elderly gentleman of Pickwickian proportions', wearing strange spectacles and a fortune in jewellery on his shirt and fingers. Lawson overheard the Pole complain because the stewards could not find him an egg for breakfast. Lawson liked him but suspected that he was enough of a gold digger to add to his jewels wherever he went.

Maryanski, first visiting Kalgoorlie in 1896, was the expert who saw the tellurides as the portent to a glorious future. Inspecting the tellurides in what was known as the 'jewel chamber' of the Kalgurli mine, he enthusiastically said that the presence of tellurides was proof that the rich ore of Kalgoorlie would descend far into the earth. 'Bunkum', replied one London engineer, but at the height of a mining boom the voices of the doubters are drowned out.

The increasing dividends of the big mines seemed to confirm the exciting news of the gold tellurides. Great Boulder, the gem of the field, increased its output of gold in almost every month. It brought more stamp mills and other machines to the field, employed more and more miners, and sank more shafts on its long strip of territory. In 1895, its first year of production, the mine's gold was worth more than £100 000; the next year it exceeded £200 000; and a year later its output exceeded £300 000. All this wealth had come from a subscribed capital of little more than £30 000. In one year the price of Great Boulder shares jumped from 23 to 200 shillings.

Just to the east of the smoking chimneys of the Great Boulder was London-owned Lake View Consols, another of the mines originally pegged by Sam Pearce. It was the next sensation. At shallow depth, in slate, it found a wonderful deposit of gold called the Duck Pond. One of the most dazzling strips of gold-bearing rock ever seen in Australia, it is said to have produced a ton of gold every month for half a year. For the year 1898, Lake View Consols passed Great Boulder in output of gold. At the end of that year it was computed that the two big mines of Great Boulder and Lake View Consols had produced half of the gold so far won from the field. Another four mines — big by Victorian standards — were also paying dividends: Hannan's Brownhill, Ivanhoe Gold Corporation, Associated Gold Mines of W.A. and Golden Horseshoe. Those who only two years ago were calling Kalgoorlie an imposter had to eat their words.

Lake View Consols rose like a balloon and was to fall like an unopened parachute. Nobody in 1898 could foresee its fate. The mine was controlled by a north of England financier, Whittaker Wright, who exploited his inside knowledge to rig the share market. Even after the Duck Pond had been drained of gold, he thought the mine was rich enough to justify his buying more shares. He did not realise that the mine's manager in Kalgoorlie was sending him misleading information about deep discoveries that had not taken place. Wright borrowed heavily to buy more shares but the price of shares rightly fell. Wright not only collapsed financially but was also convicted, in 1904, of issuing fraudulent balance sheets — fraud on a monstrous scale. Sentenced to seven years in prison, he did not even leave the court room: he swallowed cyanide and died.

For Western Australia and for Kalgoorlie the year of the Duck Pond, 1898, was magical. For the first time Western Australia became the premier producer of gold at a time when there were three prized titles in Australia — the premier producer of wool, of gold, and of manufactured goods. Kalgoorlie was supplying more than 40 per cent of that gold, and its proportion was rising. Kalgoorlie was already the premier goldfield in Australia, having passed Charters Towers, Bendigo and Mount Morgan in the preceding year. But this was only the beginning of Kalgoorlie's new place in the book of records. Several years later the output of Kalgoorlie in a single year was higher than the best year of Bendigo and Ballarat in the golden 1850s. The town's population passed 10 000, sprinted past 20 000 and was approaching 25 000 people in 1899.

THE GOLDEN MILE

The new, enlarged stamp mill at the Lake View Consols is ready to begin crushing the ore from one of the richest mines on the field. It is 1897 and every bystander wears a wide-brimmed hat. The cloth-capped man, in the front right, probably works indoors. (Photo from Battye Library, no. 10213 P)

36

The tiny Adelaide syndicate which had set this in motion was still reaping the rewards. Any one of those ten original members who had clung to his shares was now probably a millionaire. For an initial payment of £15 and maybe additional calls which were less than £100, an early-bird might now hold assets which, added to the dividends so far received, were worth about one million pounds. The exact sum cannot be reckoned with complete accuracy, but only the Broken Hill silverfield and Mount Morgan goldfield could produce stories (and only a few) to match such a financial success. Curiously, several of these Adelaide early-birds had probably not even seen Kalgoorlie.

The vehicle of their success, the Coolgardie Prospecting and Mining Co, had become so rich that it was unwieldy. In August 1897 its shareholders decided to put the company into voluntary liquidation and distribute its assets among themselves. The assets were diverse, and the liquidation was still being carried out a year later. In October 1898, close to its end, the company held huge parcels of shares in four of the six biggest mines in Kalgoorlie and in four of the lesser mines. At the current share prices the company's holding in Lake View Consols and Associated Gold Mines of W.A. was worth more than £4 million, and in Ivanhoe and Great Boulder it was worth more than £3 million. Add all the other shares still held in Kalgoorlie mines, the value of shares handed over to existing shareholders, and the £950 000 of the dividends paid by those mines, and the final figure was startling. In sum total the Coolgardie syndicate had earned assets and dividends of £13 647 000. Here was probably the most successful investment in the first century and a half of Australian economic history.

In 1900 a real estate agent passing through the wealthier suburbs of Adelaide could point to maybe ten or fifteen mansions financed from Kalgoorlie dividends or share transactions. George Brookman, the stockbroker who had founded the syndicate in 1893, was perhaps the best known of the magnates created by Kalgoorlie. He continued to live in Adelaide, presided over the board governing the Adelaide hospital, sat in parliament, and donated a small fortune to the local School of Mines. At a host of charitable functions Brookman's balding head and brush-like moustache and butterfly collar could be seen. From the late 1890s he took little part in directing Kalgoorlie mines, control of which had passed to London.

His younger brother, Will the prospector, was briefly a financial star. After leaving the camp at Kalgoorlie he went to London to help float various companies. In December 1896 he sat on the board of 21 Kalgoorlie companies based in London. According to his story they unanimously requested him to return to Australia to supervise their companies' affairs, but it was probably what is now called 'a brush off'. He travelled to Kalgoorlie in a special sleeping car hired from the railways, and was driven stylishly around the mining field in a six-in-hand, the equine equivalent of a Mercedes Benz. Everywhere he went he engaged in a little boasting and prophesying. Settling in Perth he lived in style in his mansion Great Boulder, becoming the mayor of Perth at the age of 40 and a member of the legislative council.

But Will Brookman had not learned from his earlier bankruptcy. He must have gambled heavily in gold shares. His debts soared, overtaking his assets, and he lost virtually everything, even his wife. When he died aged 50 his estate in Western Australia and in his native South Australia was valued for probate at 134 pounds and 10 shillings. A dashing and reckless figure, he had come full circle.

The goldfield was the stage for hundreds of these personal dramas, the making of fortunes and the losing of fortunes, as the shares rose and fell. Inside the mines was another kind of drama. Serious accidents were common. While these accidents, in proportion to each 100 000 tons of ore removed, were fewer than in Victoria they were still too frequent. The sheer scale of operations and the fact that thousands of men were employed meant that every year had its toll of casualties.

The inspector of mines kept his own rough and ready list of accidents, and it was probably incomplete. In the year 1898 he recorded his first accident on 10 February: three miners working on the Lake View Extended suffered shock when, owing to a faulty brake in the hoisting engine, the cage in which they were travelling stopped with a thud. Three days later in the Croesus Proprietary a miner named John Le Maier set fire to a shot in order to blast away some rock. He had climbed into a big bucket and was being hurriedly hauled up the shaft to safety when he fell from the bucket to his death. Three days later, in the small Kalgu mine, William Widdop and Jack McKnight were killed when dynamite exploded in the internal shaft in which they were breaking rock. In the following four days one man was bruised and two were badly injured. Then the reported accidents ceased for 15 days but on 8 March they resumed. Joseph Dower, working on the Golden Link leases, fell 130 feet down the shaft and was found dead and battered at the bottom. Six days later in the Hannan's Star, Eli Bull was working at the bottom of the shaft when he was killed by a lump of rock falling from a bucket high above. There were two deaths in May and one in June, but not even a reported injury in July. So the year went its way, ending with 15 deaths.

From time to time, almost-miraculous escapes were reported. Tom Jackson was working in the Great Boulder when a ton of stone fell on top of him. 'Bruises on his body' were the only damage according to the government's inspector of mines. A small mine was more likely to experience accidents than a big one. For that year Great Boulder is listed as suffering only one death and one major injury, but various small injuries must have hit its big workforce.

Still the gold poured out, in larger quantities than any single field in Australia had produced in a year. While

UNLOCKING THE JEWEL CHAMBERS

Death was all too common in the mines, but Kalgoorlie and Boulder came to treat death with all the dignity and even grandeur they could muster. This funeral held on 16 July 1907 was almost to be envied — a brass band playing and the bass-drum decked in black ribbon, the members of friendly societies or freemasons wearing their regalia and white aprons, and marching in front of the hearse. The funeral — of Arthur King, conductor of the Apollo Orchestra — was proceeding along Hannan Street.

almost every expert agreed that Kalgoorlie's lodes were rich, the quandary was how to extract all the riches. The ore was not so easily treated as in the typical mine of eastern Australia. Indeed in eastern Australia there was no big mining field — with the possible exception of Stawell in Victoria and Queensland's Mount Morgan — which so perplexed the veritable horde of resident and visiting metallurgists.

The men from eastern Australia who designed the early treatment plants for Kalgoorlie had begun with the old Bendigo formula: crush the rock in a stamp mill; collect the fine grains of gold partly by gravity (for gold was heavier and fell to the bottom) and partly with the aid of mercury which, having an affinity with gold, easily trapped it. At Kalgoorlie this process was not even very effective on the shallower oxidised ore, which normally was the easiest to treat.

The difficulty grew as the mines went down. Most ore now came from a depth of more than 200 feet; and these new masses of sulphide ore were the basis of long-term mining. When the treatment plants began to deal with this less tameable sulphide ore, they found that much of the gold was intergrown with iron pyrite. Ore of that complexity was always a challenge for the old Bendigo-type plant. Moreover, much of the native gold consisted of the tiniest of particles, and they were not easily collected. The problems of how to extract the gold were aggravated by the presence of tellurides which required new steps in the process. Normally they had to be roasted and thus decomposed, allowing their gold to be converted to a metallic state. Obviously, Kalgoorlie needed its own ingenious chain of processes for treating the ore. Until they were found and perfected, a fortune in gold would be lost each month in the treatment plants.

A Scottish invention, the MacArthur-Forrest cyanide process, provided part of the solution. First tried in 1888 at a goldfield in north Queensland, the troubled Ravenswood field, it relied on the capacity of a dilute solution of potassium cyanide to penetrate slowly the crushed material containing the gold. Eventually the gold was dissolved, after which the gold-cyanide solution flowed through a trap of zinc shavings, and there the zinc precipitated out the gold. By the time Kalgoorlie was becoming busy the MacArthur-Forrest cyanide process had been much improved. The cyanide solution did not call for the expense of heating and so was relatively cheap once the vats had been built and the cyanide bought. Above all, the cyanide ate its way into many ores that had previously been defiant. In 1896 it proved its worth at the Hannan's Brown Hill mine at Kalgoorlie, and a year later half of the cyanide vats erected in Western Australia were to be found in Kalgoorlie.

Cyanide did wonders for Kalgoorlie but it was only part of the answer. A lot of gold, perhaps 20 per cent in some treatment plants, still defied capture. The London shareholders were rarely informed of the exact loss of gold. Had

UNLOCKING THE JEWEL CHAMBERS

Men of the Lake View Consols are carrying a heavy wooden box lined with zinc to the treatment plant where it will help to trap the gold. By the law of averages 14 of these 15 strong men must have come from the other side of Australia. The year is 1896, and merely five years previously they would not have dreamed that soon they would be working in the arid country of Western Australia.

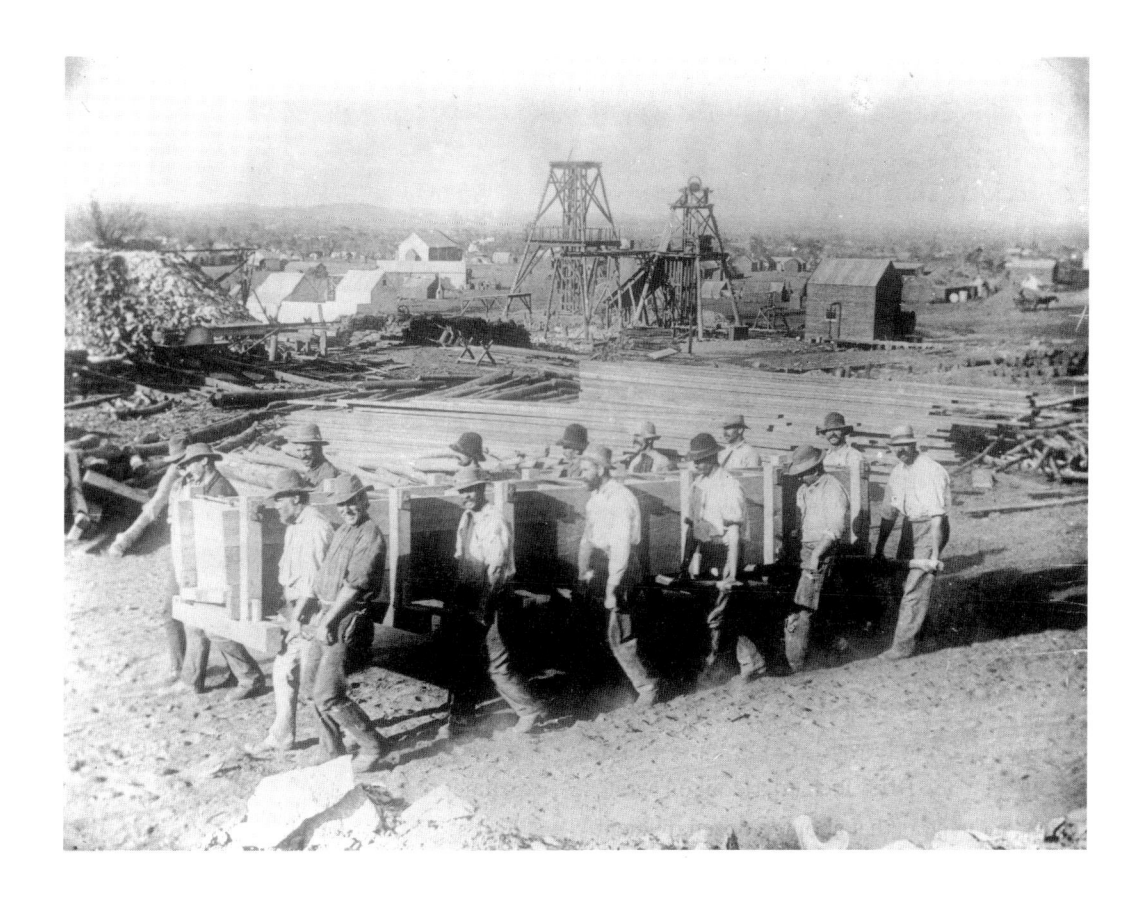

they been told, they would have clamoured for heads to roll.

Great Boulder's metallurgists were among the first to report their frustrations. Their plant began with the old Bendigo formula because their consultant, Zebina Lane, and their first general manager, Richard Hamilton, had worked in Bendigo and knew well its processes. But the Bendigo formula was inappropriate: by the end of 1896 the treatment plant at Great Boulder was almost surrounded by growing dumps of waste residues and slimes rich in gold. In some weeks more gold escaped than was captured by the expensive use of the old mercury and the new cyanide. It was almost as if the treatment plant had diarrhoea: much of the gold flowed out with the waste.

It was recognised that in the long term many of the big mines would be worthless without a new way of extracting nearly all the gold in their lower-grade ore. New inventors had to be found. They came forward, in person or in the daily post. On 17 January 1897 Richard Hamilton wrote to the Great Boulder directors that about 'a dozen different processes have been placed before me, all perfect successes; at least the owners state so'. Consultants were called in but some of their advice was contradictory. In the end Great Boulder was tempted to build roasting furnaces and adopt the method of treatment said to have been successful with telluride gold ores at Cripple Creek in Colorado. Zebina Lane swore the roasting process would also be frugal in using water. So in 1898 the roasters were erected at the mill in order to eliminate as much sulphur as possible from the ore and also to help release the imprisoned gold.

The sulphurous smell of the Great Boulder's roaster plant pervaded that end of the field. While the town dwellers got the aroma, the foreign owners of the patent got, in the form of royalties, much of the gold extracted. Great Boulder soon realised that it was paying dearly for a process that still did not capture a high percentage of the gold embedded in the rock. The shareholders, meeting in Winchester House in Old Broad Street, London, began to rebel. The contract with the owners of the roasting patent and process was torn up. The machinery now discarded at the mine could not so easily be torn up. In years to come the visitors who knew their metallurgy were surprised to see little-used equipment, imported from England and Germany and South Australia, gathering red dust in corners of the leases. That was the price of the ceaseless experimenting.

All the treatment plants, in crushing or in pulping the rock, produced too much of what mill men called 'slimes'. These fine grains were smaller than 1/400th of an inch in diameter. They were so fine, so reluctant to settle down in a tank of water, that it was almost impossible to separate each grain of gold from the innumerable grains of barren rock. Slimes did not lend themselves easily to the valuable cyanide process, increasingly used to extract gold. In the words of Donald Clark, slimes were a 'terrible bug-bear'

UNLOCKING THE JEWEL CHAMBERS

One of the Lake View Consols' first filter presses — a Kalgoorlie machine that added much to the wealth of Australia.

for all old-time metallurgists. Then almost out of the blue a young man at the Lake View Consols contrived a simple process for dealing with slimes — so simple it could only be found with a touch of brilliance.

The brilliant metallurgist, John Sutherland, was trained at the Ballarat School of Mines, and first worked in Broken Hill. A quiet, reflective bachelor, he realised early in 1897 the relevance of the filter press for the difficulties created by the slimes in his treatment plant at Kalgoorlie. The filter press was used in sewage works for separating water and solid waste, and in sugar-crushing mills for squeezing excess water from pulped sugar beet. Obviously, with adaptation, it could be used for percolating water from gold-bearing slimes. In effect the steady application of pressure forced the dandy liquid through canvas or strong cloth, leaving the muddy slime behind. A big mill would need a long row of filter presses to treat a big tonnage of slimes each week but it could be done. The filtered slimes were more amenable to the new cyanide process, and so most of the fine gold was recovered.

Sutherland placed his ideas before the general manager, Charles Kaufman, who gave permission to spend money on a trial plant. It worked. Sutherland was then asked how much money he would need to install a large-scale plant in which all slimes could be treated in filter presses. Sutherland gave his figure as £5000, then a largish sum. 'Well, boy,' Kaufman is said to have replied, 'there is £10 000 placed at your disposal. Let us see what you can do.' Sutherland's process was eventually to be used in gold mines around the world. Later the distinguished American engineer Herbert Hoover claimed to have devised the process at Hannan's Brownhill mine at Kalgoorlie but his claim is hard to take seriously. Certainly, Hoover and Hannan's Brownhill were early advocates of the filter press but Hoover had not even reached Kalgoorlie when Sutherland began to adapt the filter press.

The slimes produced by the thundering lines of stampers in the treatment plants were no longer a pest. Now that the slimes could be treated and dried by the filter press and prepared for roasting, there was actually an incentive to use crushing methods which produced a larger proportion of slime. As finely-ground ore was likely to produce a larger share of slimes, fine-grinding machines came into favour. A German, Dr Ludwig Diehl, went a step further at Kalgoorlie. Representing the London & Hamburg Gold Recovery Co., which was floated on the London stock exchange in 1896, he borrowed another device from another industry and applied it to the milling of gold. He adapted the tube mill that was used by the cement-making industry for the crushing of limestone. At Hannan's Brownhill at Kalgoorlie, where his company had a contract, Dr Diehl and the James brothers set up what was to be the mother of tube mills for the world's gold industry. Their tube mill was a cylinder filled with water and hard pebbles. It gave a second crushing to broken ore which had already been pounded by the stampers. As the

UNLOCKING THE JEWEL CHAMBERS

Treating gold-bearing ore in the Hainault mill. Several vertical iron stampers can be seen in the stamp mill on the left, while the grinding pan is at work on the right.

45

tube or cylinder revolved, it slowly converted the broken gold-bearing rock into a slime or pulp. The German also used Sutherland's idea of the filter press to remove the water from the slimes and prepare the slimes for roasting. Possibly he was toying with such an idea when Sutherland made his breakthrough.

In a revolutionary sequence of treatment Diehl set out to produce in his tube mills nothing but slime. The water was then squeezed from the slimes in a filter press — he used a German model known as the Dehne press. His next step was to find a better form of cyanide mixture for extracting the finer grains of gold. He selected a combination of the common cyanide of potassium along with the bromide of cynogen first used in Canada by Sulman and Teed. From 1899, the Diehl process, known loosely as wet-crushing, was a landmark in the history of the field. Many mines copied it.

Another avenue of Kalgoorlie experimenting ended with a 'dry-crushing method'. In 1902, Great Boulder's version of this consisted of the following steps: Gates Crusher or stamp mill, Griffin mill, Edwards' roasting furnace, amalgamating pan and *spitzkasten* or classifier. The remaining slimes were sent to the big cyanide vats, and so to the filter presses and gold-smelting rooms. Versions of this dry-crushing process were the ultimate winner in the battle of the techniques.

Visitors who knew the metallurgy of a typical gold mill in Bendigo saw, when they reached Kalgoorlie, sights and sounds and processes that surprised them. By 1900 the stamp mill and old amalgamation process were no longer the main way of extracting the gold. Whereas the thump of the iron stampers was still to be heard in every big gold mine in eastern Australia, the stampers were being discarded in a few of the big Kalgoorlie mines. The result was less noise, but an increase in smoke because the roasting of all or part of the crushed ore became part of the sequence of treatment processes. On still days a white blanket of smoke hung over the streets near the mines. Another unusual sight at Kalgoorlie was the filter presses in all the treatment plants. Visitors with any curiosity wanted to see this special Australian invention quietly doing its work, and no man was pointed out with more pride than young and thoughtful Mr Sutherland, who was even said to look like an inventor.

The gold towns, like the mines, were about to pass beyond the experimental stage. Kalgoorlie in 1898 still had the appearance, from certain angles, of being blown there by the wind and dropped onto the ground. Thousands of tiny huts and tent-houses and pretend-houses stood on their allotment with hardly a fence between them, rarely a tree to shade them and the surrounding ground almost bare, as if swept by the winds. Many families from Victoria had been told what to expect but some were dismayed when they reached Kalgoorlie. Ernie Carter, aged 5, stood on the railway station with his two brothers and two sisters, and even when he was 90 he remembered clearly that desolate Sunday in 1899: 'it was hot and dusty and I can recall my poor old Mum, she cried on the platform.' From the railway station the newcomers walked or if they had a little money they went in a horse-drawn cab or cart to their first house, usually close to the mines.

Most houses were makeshift, and few were built of brick or weatherboard. Many had no floorboards, just the sand to walk on. Eventually the sand was covered — as a first measure — by strips of coarse cloth scavenged from the filter presses and scrubbed tolerably clean so that they formed a hard mat on the smooth sand. As most houses lacked a wash-house, the washing of clothes took place in the open air. A fire was lit, water was heated in a kerosene tin, and the dusty clothes were often beaten with a stick before being washed: at least they dried speedily in the burning sun. In many simple kitchens the original stove was homemade, while early items of furniture were nailed or screwed together from packing cases and scavenged lengths of timber. After a few months of earning high wages the miners began to improve their houses, often bringing — at high cost — their old furniture from the eastern states.

The goldfield now supported two large towns, not one. Boulder town, close to the rich mines at the south end of the field, was born in December 1896. Its first blocks of land were sold to shopkeepers and especially to miners who previously had made themselves 'a nuisance' by camping on the mining leases and obstructing the development of the mine. Next year Boulder gained its own municipal council and a busy railway line to Kalgoorlie. By the first years of the twentieth century Boulder held over 5000 people and was informally calling itself Boulder City. From Boulder to Kalgoorlie stretched an unbroken line of mines, treatment plants, mine dumps, huts, houses and shops, churches, halls and hotels. By 1905 Boulder held at least 26 hotels — more than were left in waning Coolgardie — as well as three banks which on payday sent to the mine offices the gold sovereigns which many miners greatly preferred to banknotes.

Boulder, while booming, was the poor sister. Kalgoorlie was still the commercial hub, with six banks and several hundred shops, three breweries and five hospitals, and a busy railway station from which lines by 1905 extended in four different directions (to Perth, Boulder, Kanowna and Menzies) and horse teams and strings of camels carrying goods where the railway would never extend. Kalgoorlie was also the goldfields' centre of culture and entertainment, with eight denominations ranging from Catholic to Church of Christ and Congregational and Salvation Army. The red-light capital of Western Australia, its dozens of French and Japanese prostitutes catered for a population in which men far outnumbered women, though the ratio was falling. Curiously it was the only major town in Australia where Japanese were more numerous than the Chinese: as late as the year 1907, 16 of the 21 licensed laundries were run by Japanese. Kalgoorlie was also the town of printers and newsagents with its penny daily paper the *Kalgoorlie Miner*, its weekly *Western Argus*, and its Sunday *Sun*, supplemented by an afternoon daily, Boulder's *Evening Star*. In the House of Representatives of the first federal parliament, two of the only five seats allocated to Western Australia were won by men who edited Kalgoorlie newspapers. John Kirwan of the *Miner* and Hugh Mahon of the *Sun*.

As confidence grew in the future of Kalgoorlie, grand buildings arose, some built with the charming pastel local stone and many with bricks of delicate red. Wide Hannan Street imitated the grandeur of Sturt Street in Ballarat with all its ornamentation but without the advantages of its statues and leafy English trees. Some of the finest Australian streetscapes that survive from the decade 1898–1908 are in Kalgoorlie, and it would be one of the country's busiest tourist resorts today if it happened to lie within a few hours' drive from Sydney. The Palace Hotel and its adjacent arcade, opened five years after Hannan's discovery of gold, was one of the most imposing hotels in the continent; and soon it was bought by an overseas syndicate and floated on the London stock exchange where it announced its annual dividends of 10 per cent.

On the veranda of this hotel the author May Vivienne sat one afternoon in 1900 and jotted down her surprise at the rich clothes of the fashionable women strolling the pavements, including gowns which she understood were ordered from Paris and London: 'Occasionally a plainly-dressed woman in a tweed or Assam silk costume with neat sailor hat would pass, probably a mine manager's wife or English visitor, but the majority of the women of the goldfields spare no expense in the style or riches of their dresses.' No doubt she exaggerated: she did not have time to visit the shopping streets of Boulder and observe dowdy wives, children at their feet, counting every halfpenny they slipped into their purse.

May Vivienne pluckily went down gold mines, shuddering when she realised the dangers faced by miners but also imagining, when rich ore was pointed out to her, that she was at the entrance to Aladdin's Cave. At the 800-feet level of the Great Boulder she was delighted to see the miners in the candlelight sitting on big stones and eating their midday meal from their individual shining dinner-cans consisting of three compartments — for tea, for bread and meat, and for sweets. Above ground she saw the men laying the electric tramways along the main streets for a London company which hoped to make a fortune from the thriving goldfield. And in the dining room of a hotel, beneath the whirring fans, she heard starry-eyed conversations which she repeated in her handsome book *Travels in Western Australia*. Kalgoorlie, she proclaimed, might one day have 300 000 people.

In the first years of this century the combined metropolis of Kalgoorlie-Boulder was one of the ten large metropolitan areas of Australia. The cities of Sydney, Melbourne, Adelaide and Brisbane formed the big four, and then came

UNLOCKING THE JEWEL CHAMBERS

Part of the township of Boulder stretches away in the background. In the foreground is the Ivanhoe South Extended mine, 1905.

Newcastle, Perth, Ballarat and Bendigo to form the next four. Hobart followed with 35 000 people, while Kalgoorlie-Boulder between them held well over 30 000. These two golden towns, virtually forming one elongated city, together held tenth place in Australia but in terms of contribution to national wealth they were probably eighth.

By the year 1900 the catchy term used increasingly to describe the heart of the mining belt was 'The Golden Mile'. This arresting phrase covered only the southern end, around the Great Boulder and the Ivanhoe, Perseverance and Lake View mines. Significantly, the smaller mines at the Kalgoorlie end — the site of Paddy Hannan's original discovery — were not even counted as part of the Golden Mile. The term Golden Mile was really a charming understatement, a sign of the quiet confidence of this extraordinary metropolis. After years of boasting prematurely about its hidden wealth, Kalgoorlie was content to call itself a mere mile of gold when in fact its producing mines straddled a north-south line for at least four miles.

Nothing showed more vividly the spirit, attitudes and riches of the rising goldfield than its absorption in sport, especially games that involved a gamble. Boulder and Kalgoorlie each had its racecourse with fine grandstands, terraces and enclosures where finely-dressed women and men could promenade. For the horses' owners the prize money was high, and the Kalgoorlie Cup came to rival the Perth Cup as the main horse race in the state. The most celebrated Cup-week in Kalgoorlie's history was in August 1904. The governor Sir Frederick Bedford and his entourage spent Cup-eve at the art *conversazione* in the Caledonian Hall where the local guests, mostly in evening dress, were so jammed as to prevent him from approaching closely to many of the works of art on display, though at least he heard loud and clear the Boulder Orchestral Society play a waltz composed by its own conductor. Next day from the grandstand the governor saw a brown horse named Blue Spec win the Kalgoorlie Cup. He was to hear more of Blue Spec because next year it won the Melbourne Cup, beating the record set by the lightning horse Carbine.

In that era the links were close between mining and racing, and the winner of the following year's Melbourne Cup was Poseidon. That horse was such a favourite that its name was borrowed for the christening of new gold-mining companies and new-found gold nuggets. More than 80 years later, the company Normandy Poseidon — indirectly honouring that famous horse — was to acquire the controlling interest in the entire Kalgoorlie goldfield.

Meanwhile on the goldfields the Crimson Flash became as well known as Blue Spec. Crimson Flash was the nickname of Arthur Postle, a professional foot-runner who came from the Darling Downs. Often he ran at Kalgoorlie when it was becoming one of the world's few centres of foot-running. The fifth of December 1906 was memorable in the calendar of Australian sporting history, for in Kalgoorlie the world champion sprinter R. B. Day of Ireland was set to defend his title. The world championship con-

UNLOCKING THE JEWEL CHAMBERS

A stretch of the Golden Mile in 1905. This uncrowded section shows chimneys, poppet heads and white tailings dumps in the distance, and of course the railway lines, treatment plants, miners' huts and managers' houses, many stacks of firewood (see bottom-left corner) and grass-free earth. Hardly a wisp of smoke is visible. Almost certainly the photographer has decided to do his work on Sunday, climbing to the top of the Ivanhoe poppet head and directing his camera towards the Great Boulder in the middle distance.

THE GOLDEN MILE

In 1907 the meetings at the Kalgoorlie and Boulder racecourses had all the style of city meetings.

UNLOCKING THE JEWEL CHAMBERS

Spectators gather in Tattersall's Club in Kalgoorlie on 23 May 1907 to watch local champion Fred Lindrum junior, who was the older brother of the still-unknown Walter Lindrum. Young Fred is playing the visiting English star, Mr Melbourne Inman. With one hand holding the cue and the other hand in his pocket, Fred stands on the left. If it had been a hot day the great fan would be whirring overhead. Obviously this is one of those infrequent days when rain was expected — we can see that unusual sight in an early Kalgoorlie photograph, a man (second from right) carrying a black half-folded umbrella.

sisted of three sprints over varying distances from 75 to 300 yards, and a crowd estimated at somewhere between 15 000 and 20 000 assembled in Kalgoorlie to see, to its delight, Postle's crimson colours race to the front.

In the era when professional athletics had a standing at least as high as that of the Olympic Games, some of the most discussed records in the professional book of records came from Postle's heyday. At the small gold town of Menzies, on the railway line beyond Kalgoorlie, Easter Monday of 1906 was listed in the professional record-book for several decades to come because Postle ran the 130 yards — then the popular professional distance — in the world record time of 12 seconds. In December 1908, back in Kalgoorlie, he ran 100 yards in the world record time of nine and a half seconds. Both records were disputed and disallowed because the Kalgoorlie and Menzies sprint tracks had falls of about one yard. It was typical of the goldfields that so many sports activities were slightly hit-or-miss but still world class.

Billiards, like professional running, attracted a torrent of wagers and bets. Hotels on the goldfields with any claim to class had their billiard room with a manager and sometimes an attendant, presiding full-time and promoting gambling on the billiard games as well as serving as bookmakers on horse races at Melbourne and other far-off courses. The best-known billiards family on the goldfields came from Port Melbourne in about 1896, appearing first at Coolgardie and then renting the fine billiard tables installed at the grandest institution in Kalgoorlie, the Palace Hotel. The head of the family moved to the green-cloth tables at the nearby Australia Hotel in 1899 before adjourning to the coast, returning seven years later to preside over the billiard tables at the Great Boulder Hotel in Kalgoorlie and then at one of the seven hotels at Broad Arrow, a shortly-shining gold town 24 miles to the north west. It was during the family's stay at the Palace Hotel in Kalgoorlie in August 1898 that one of the sons was born. His initials were W.A., for he was the first of the family to be born in the west. His full name was Walter Albert Lindrum, and he was to become the best billiard player the world had seen.

Billiards was one of the many gambles that supplemented, on the vigorous goldfields, the core of the gambling craze — the search for gold. Some players were also attracted to billiards, they said, because the table was the only green sward they regularly saw in that most arid of districts — until the year when a stream of fresh water magically arrived. How that water was secured, constituting one of the triumphs of Australian history, was itself something of a gamble.

THE TORMENT OF THE WATER KING CHAPTER FOUR

ALONG THE GOLDEN MILE in 1902 the event talked about incessantly was the water scheme. A pipeline was slowly approaching from the coast, a long snake of steel pipes large enough for a child to crawl through. One of the most remarkable public works the world had seen, it was probably longer than any big pipeline previously built in the world. It was also controversial.

Charles Y. O'Connor designed the goldfields' water scheme. In 1902 he was aged 59 and at the peak of his career, though not enjoying his work. After gaining experience of railways in his native Ireland he migrated to New Zealand, where he made his reputation in the South Island, designing roads across the alps, improving harbours and building railways. In the towns where he lived with his large family — in Christchurch, Hokitika, Dunedin and Wellington — he was a commanding figure, lean, tall and athletic, witty and pithy with his words, and wide in his sympathies. Disappointed that he had not been made head of the public works department in New Zealand, he was tempted by Sir John Forrest, Premier of Western Australia, to move to Perth in 1891. When he enquired of Forrest whether his duties as the government's chief engineer embraced roads or railways or ports, he received a telegram saying 'Everything'.

For a few years O'Connor took on everything, supervising the railways, designing the new harbour at Fremantle, and providing water for dozens of arid goldfields or the lonely camping places on the long inland tracks. He held what was probably the most creative engineering job in all Australia. O'Connor was also that rarity, an Australian engineer honoured by Queen Victoria, for he had been appointed Companion of St Michael and St George five years previously.

In 1902 his bold water scheme was within a year of completion. The trickiest stretch of the pipeline had just been tested and found to be sound. Outwardly he seemed to be on top of the world. Inwardly he was in a state of tension, even torment. He was suffering the pain of neuralgia, burdened by his own worries and those of his work. Irritable in the day, he was sleepless at night.

It was his practice early each morning to ride his horse from his Fremantle house, often passing the harbour, his own harbour, where the big English, French and German oceanliners now berthed. On the morning of 10 March 1902 he carried a revolver. Riding beyond the harbour and southwards along the sandy shore of the Indian Ocean, he then turned his horse into the shallow sea and shot himself.

The telegraph carried the news to the goldfields where the shock — 'sensation' was the journalists' word — was intense. People living there even speculated that O'Connor must have known something unfavourable about the water scheme, and that therefore the success of the pipeline itself could be in jeopardy at the very time when O'Connor claimed it was close to success. These fears and rumours were not altogether quelled by the fact that he left behind a note expressing his faith in the water scheme and his anger at the public attacks on his work.

A year later the fresh water reached the goldfields. The rehabilitation of O'Connor's public reputation slowly began. Long after his death he was becoming a folk hero. Perhaps no public servant in the history of Australia is so venerated. He is now seen as the engineering genius who thought boldly but was hounded to death by the criticisms of petty or unimaginative opponents. He is now venerated as the engineer who did the impossible, bringing water to the far desert and enabling a goldfields city to arise — though in fact the city was flourishing before the water arrived. His critics, in contrast, are seen as mean and blinkered, the enemies of progress. The truth is not as simple as that.

The idea of finding a special supply of water for the parched goldfields had become a popular topic of debate in 1895. The mines were booming, the population was growing, and water was scarce. It was in late 1895, at the start of another long summer, that Sir John Forrest gave public hints about the water scarcity. He was then visiting the goldfields for the first time in eighteen months. It seems incredible that he should for so long have been out of personal touch with the region that was becoming the dynamo of his colony and the indirect source of most of its revenue. But he was a busy man — in effect he was the government — and it was difficult to find time for the long journey to Coolgardie and Kalgoorlie. Nor could anybody accuse him, the most famous explorer of his own territory, of not knowing the outback and its droughts. However, there was another reason why he did not visit the goldfields while they were surging ahead: he did not depend on the goldfields' votes. He saw to it that the goldfields, so far, had few votes!

In Coolgardie at last, on 22 November, Sir John listened to deputations. One of their demands was for a secure supply of water. As his government had collected the large sum of £100 000 from the sale of town sites in Coolgardie alone, perhaps he might think of spending part of that money on water? A sensible politician, Sir John nodded, saying neither too much nor too little. While being driven in a horse-drawn vehicle through the dust to Kalgoorlie, he was informed that many of the promising mines were temporarily not working: they were short of either water or men.

On Sunday 24 November, Sir John was another twelve miles away, at White Feather — the present ghost town of Kanowna — where the committee of the local stock exchange welcomed him. His mind was moving towards the water question, and in the evening he would have the chance to give his views, because about 40 people — including at least one woman — were to honour him at a banquet in Donnellan's Criterion Hotel. In many ways it was a night to remember. There were even flowers on the tables — they must have been wild flowers — and pieces of fruit conveyed by wagon and then railway train and express cart all the way from the coast. Mr Donnellan

THE TORMENT OF THE WATER KING

In 1902, a year before the arrival of the water pipeline from the coast, the inhabitants of most houses and huts around the Boulder mines had to buy their water from a travelling cart or instal a small iron tank in which to catch rain that fell on the iron roof. In this desultory suburb near the famous Ivanhoe mine, every house seems to have its isolated lavatory or outhouse but only the bigger houses own a water tank.

must have been delighted to open his copy of the *Coolgardie Miner* about a week later and find that his 'viands and the general excellence of the culinary arrangements left nothing to be desired'.

The seating at the top table was carefully arranged, with the major of Kalgoorlie, J. Wilson, sitting close to Sir John. There were loyal toasts and the singing of 'For He's a Jolly Good Fellow' and a host of requests for the Premier to consider: a school, a daily mail delivery from Kalgoorlie, a smoother road from Kalgoorlie, and perhaps even a branch railway when the time came. And there was the question of the fluid which that night probably was not prominent on the loaded tables — water.

While scarcity of water was not the dominant topic in the eyes of the mine managers, businessmen and officials present, the local member, C.J. Moran, hoped the government would sink bores which yielded salt water for the condenser men to convert to fresh water. One mine manager, Nat Harper of the Robinson gold mine, was much bolder. He suggested that a water pipeline be laid all the way from the ranges near the distant town of York. He could not have suspected that his idea that evening would be taken up by the Premier.

Forrest, toasted and warmly cheered, was bluff and hearty, expressing his pleasure at seeing several Western Australians in this gathering of people from 't'other side'. The old fox added that he was delighted to be a guest 'in this promising and important township'. The shanty town of White Feather had never been so flattered. When he turned to the water question he was equally foxy, for he knew that he must let slip past his lips no promise that would live to haunt him. He must promise no more than he could fulfil. Accordingly, he said that one of his tasks was to find a way of ending the scarcity of water on the goldfields. The eternal drought 'will, I believe', be overcome. How he would overcome the drought he did not yet know, but he said that if the goldfields proved to be as permanent as was believed, he would even consider pumping water from the coast — if a pumping scheme would pay. 'What a transformation there would be if the district had a stream of fresh water running through it,' he said.

Sir John Forrest was the political author of the rather vague plan to bring water to the parched goldfields. If he went ahead he would have to find the money, a frighteningly large sum. And he would have to take the political risks — even higher than the engineering risks. If the scheme failed, or if it worked as an engineering venture but proved to be a life-long burden to the taxpayers, he would have to carry the blame on his broad shoulders. If the water scheme failed, so huge was the money at stake that his political career would probably be at an end. So while O'Connor's suicide served to turn him into a legend, Forrest was at least as much the hero for the bold attempt to bring water to the goldfields.

Once the water pipeline was almost on the political

agenda, Charles O'Connor's advice was crucial. No doubt he had already spoken to Forrest about the costs and difficulties. As the goldfields pipeline was discussed more and more, O'Connor was increasingly its quiet advocate and architect. He quickly — almost too quickly — concluded that the long pipeline was the only way of providing Forrest with his 'stream of fresh water', but he soon glimpsed the troubles kindled by the scheme. After he reported that a long pipeline from the wet coastal ranges to the dry goldfields was the only sound solution, he was denounced by critics as the inventor of the water scheme. It was disparaged as O'Connor's 'fancy project'. Here, said critics with pointing finger, was an official engineer using the public funds as a toy with which to perform grand experiments.

In his official report in July 1896 explaining his pipeline to the government, O'Connor revealed his touchiness: 'I never urged, nor do I now propose to urge upon the Government or the country, the undertaking of this work.' He emphasised that he was simply the man of whom a question had been asked. His reply to that question was that if the government wanted more water for the new gold region of which Coolgardie and Kalgoorlie were the heart, it would be cheaper to build a reservoir near Perth and to pump the water all the way from the coastal ranges to the distant goldfields.

People were entitled to marvel or snipe at his solution. It was audacious, but the audacity is now lost to us because we are accustomed to such projects. In 1900, however, water had probably never been moved so far. In Italy the Roman engineers had built remarkable aqueducts to carry water long distances over rugged terrain. Thus the Anio Vetus, commenced in 272 BC, carried water 43 miles from Tivoli down to Rome, from the hills to the foothills and the coastal valley. The Aqua Marcia, originating close to the Italian Subiaco, conveyed water 62 miles to Rome. In the last decades of the nineteenth century, engineers again matched these Roman projects. In 1879 the English parliament authorised an aqueduct to carry water 96 miles from Thirlmere to the city of Manchester. Of the first thirteen miles, eight were tunnels through hills and mountains. The cast iron pipes were 40 inches in diameter, compared to the 30 inches O'Connor had in mind.

However, there was one startling difference between these old Roman and new English schemes on the one hand, and the proposal that O'Connor made for Western Australia. In the other schemes the water flowed downhill. O'Connor, in contrast, would push the water uphill. Already in a few parts of the world the latest pipelines occasionally carried oil uphill — one crossed the ranges from the Caspian Sea to the Black Sea in tsarist Russia. But such pipelines were babies compared to the one O'Connor proposed. No wonder people talked, enthusiastically or sceptically. O'Connor and his plan provided kindling for arguments, above ground and below, in booming Kalgoorlie, which now was likely to be the main consumer of the water pumped from the coastal ranges.

Today the typical Australian knows that much of the west is a vast plain and that near Perth the coastal ranges are low. Accordingly, the idea of pumping water from the coast to the interior is at first sight the easiest of tasks. In fact the extended plateau on which Kalgoorlie rested was well above sea level. The approach to the goldfields was long and gentle but the rise was substantial. To reach Coolgardie, the water had to climb to 1400 feet above sea level. In fact Coolgardie was as high as Ballarat. Anyone who had walked from the ports to Ballarat in the gold-digging days knew that it was an uphill climb, so much so that not until 1890 did the railway first join Melbourne and Ballarat by the direct and shorter route, partly because the climb for locomotives was so steep beyond Bacchus Marsh. The pumping of water to the goldfields was not easy.

Kalgoorlie was about 150 feet lower than Coolgardie and so the water could flow by gravity on the final stage of the pipeline; but when O'Connor first worked out the broad details he did not necessarily envisage sending the water as far as Kalgoorlie. The hopes of Western Australia were still centred on Coolgardie: its drastic slump was not predicted.

O'Connor had to be persuasive in selling his pumping scheme. Fortunately he could write with a clarity and force that few of today's engineers and public officials would surpass. Assailed with the disbelieving question, 'you will pump water all that way?', his reply was so practical that most who listened could readily believe him. He explained that very few pumping stations would be needed. As for fuel, the pumphouses would raise steam by burning the local firewood. Nothing could be simpler.

One of his clever arguments was to point to the smallish steam vessel which each day could be seen dredging sand from the bottom of the new Fremantle harbour. *The Premier*, a new suction dredge, was capable of filling her barge or hopper with wet sand in the space of 20 minutes. O'Connor did his calculations and came up with the conclusion that the dredge, at that rate, was capable of lifting a weight in sand and water equivalent to 50 million gallons a day. The steam pumps in that one dredge, according to O'Connor, were capable of lifting to a height of 100 feet all the water that the pipeline would have to carry. In other words, a pumping capacity equal to that of about one dozen dredges would be enough to pump a continuous flow of water all the way to the goldfields. He made it sound so easy.

O'Connor had the knack of thinking up a vivid example that people could understand. He disarmed his critics with parables. When some said a pipeline would be too dear and that instead, a deep shaft should be sunk in search of water on the goldfields, O'Connor replied decisively. He said that if they chanced to find a buried pool of water at a great depth somewhere near Coolgardie, the cost of pumping it just half a mile to the surface would be even greater than the cost of pumping it some 350 miles from the coast. It was a telling argument. What he did not explain was that water pumped from a deep shaft would not need the costly 350 miles of steel pipes!

The idea of pumping water to the goldfields interested private investors. They too thought it could be profitable, if done their way. As far back as 1894 a building contractor, John Maher, had approached the government for the right to build a line all the way from the river near the town of Northam, about 60 miles from Perth. In February 1895 other promoters talked of building a line of narrow pipes — a mere 14 inches in diameter — to Coolgardie from the coast far to the south. It would be only two-thirds the distance of the O'Connor scheme; and with the thin pipe and the shorter mileage it would be relatively cheap. Where would the water be found? The promoters, eagerly questioned, said they would pump salt water from the Southern Ocean at the little port of Esperance or fresh water from lakes said to be near Fanny's Cove. In Coolgardie the salt water would be sold for 20 shillings for each thousand gallons — very dear water — and used for mining operations, or condensed at further expense into fresh water. The plan vanished with the promoters.

The Wilson brothers, who had made their fortune on the Broken Hill silverfield and were now big investors in the infant Menzies goldfield, to the north of Coolgardie, talked of sending water — presumably by gravity most of the way — to their gold leases and so on to Kalgoorlie. They called their plan the Coolgardie Goldfields Universal Water Supply Scheme. They believed they had a fine water catchment near Gum Creek, about 60 miles north of Menzies. Like all promoters they hoped the government would grant them a monopoly of supplying water on the goldfields around Kalgoorlie. To Forrest, however, their scheme was second rate. Once the government's own scheme was under way no other was permitted. The government wanted the monopoly.

On the goldfields on hot summer evenings, sitting under the stars, groups of people would find conversations turning again and again to water. They should sink artesian wells, said the Queenslanders. Others often suggested deep shafts which would surely find water, especially if the shaft reached sea level. According to present knowledge neither plan was sound but in 1896 they were widely supported. O'Connor had no patience with these schemes. There was unlikely to be artesian water, he said. The geologists told him so, and his own experience agreed. And a big mining shaft or even a borehole sunk in search of water would take years in the sinking. At his fingertips were supporting figures — some of his figures were too supporting — from the big goldfield of Charters Towers. There a deep bore drilled by the Golden Gate company was taking a very long time — progressing by a mere 30 feet a week.

Such evidence suggested to him that a bore sunk 3000 feet in search of water at Coolgardie would occupy two years. Even if it found fresh water, the thin borehole would yield each day the 'merest trickle'. So a big shaft would have to be sunk, and more years would pass. And would an underground reservoir, if found, be large enough? O'Connor warned that 'it might consequently be found, after an enormous expense had been gone to, that it would run dry in a few weeks or months'. In contrast to these plans O'Connor promised speed. Within three years he could pipe water into Coolgardie. As an advocate, as an arguer, he could not be put down.

Forrest respected O'Connor and his judgement. The two men were not frightened of bold schemes. In March 1896, when speaking at the opening of the railway at Coolgardie, Forrest firmly promised that the government would solve the water scarcity just as it had solved the transport problem. He spoke of the probability of a pipeline from the coast. Already O'Connor and his staff were busy, knowing that Forrest sometimes wanted to do things with what O'Connor called 'a mighty hurry skurry'. In September 1896, only two months after he received a copy of O'Connor's closely-argued report, Forrest introduced to the parliament in Perth a bill to spend about £2.5 million on the Coolgardie water scheme. The debate was vigorous; some said Forrest was simply offering an expensive bribe to the goldfields voters, but the bill was passed without even a formal division into the ayes and the noes.

The plan was almost in place. A dam or dams would be built across a gorge in the ranges near Perth — Mundaring weir was eventually chosen — and eight big steam-driven pumphouses would send the water 351 miles to the goldfields. To finalise the details of the pumps and pipes, O'Connor went to England for discussions with three distinguished engineers. He returned outwardly as confident as ever.

The scheme required no revolution in pumping. This was in many ways an orthodox engineering scheme: it was the scale that was so ambitious. For Forrest the financial implications must have been staggering. The large cost of the scheme can best be compared to that of building the railways. By June 1897, Western Australia had built with speed a network of inland railways that were one third as long as those of New South Wales; in relation to population Western Australia already had built the most mileage. The total cost of equipping and building these railways was £3.7 million, and so the new water venture would require two thirds of that total.

Money for such a costly project was not easily borrowed in London. Meanwhile, just two miles from Coolgardie, the government set out with the slim hope of finding artesian water at a depth of half a mile or more. If a big artesian reservoir were found far below the hot earth, the pipeline would surely be forgotten. After 16 months of drilling, the thin borehole reached 2900 feet. Not a trickle of water was found. At his big desk in Perth, O'Connor must have quietly purred with pleasure.

In 1898 the work on the weir was begun. Two large foundries — Mephan Ferguson's of Melbourne and the Hoskins brothers' pipeworks in Sydney — were selected to fabricate in Western Australia the 60 000 long steel pipes. This was the most expensive item in the whole budget, absorbing four of every ten pounds spent on the water scheme. Using imported steel, for Australia had no steelworks, they fabricated the long lengths of pipe using Mephan Ferguson's ingenious locking-bar device. In 1900 the workmen began to make the concrete wall of the Mundaring dam, said to be the highest overflow weir in the world. The manufacturing and assembling of the powerful pumps was begun. In 1901 the first of the 30-inch steel pipes were laid, mostly in shallow excavations, and joined one to another with a simple sleeve joint. So the mighty pipeline made its way eastwards, crossing creeks with the aid of narrow bridges, half buried in the earth as it skirted low hills, and finally entering the vast plain. At the end of January 1901, only 92 miles were completed.

As the pipeline was laid across the plains towards Coolgardie and Kalgoorlie, controversy followed it mile by mile. O'Connor killed himself but the work went on. Would the pipeline and eight pumping stations be efficient enough? Would the water, when at last it arrived, be too dear? These questions were pushed aside by the vigour of the builders of the pipeline. By the end of 1902 the pumping stations and their shining machinery were virtually complete. The long pipelines, the Mundaring dam near the coast, and the small holding-tanks at Coolgardie and Kalgoorlie, were ready. The pipeline had reached the Coolgardie suburb of Toorak and its extension was only a fortnight from Kalgoorlie.

Ironically the previous year had seen remarkable rains at Kalgoorlie and poor rains at Mundaring. The reservoir, which was the essence of O'Connor's scheme, held little more than 500 million gallons. This was only nine times as much water as the long pipeline itself held when full of travelling water. In fact the reservoir held barely enough water to sustain the scheme for one month if by chance it was to work at its planned capacity. Fortunately, when winter came, rain filled the reservoir at Mundaring.

At lunchtime on Thursday 22 January 1903, all was ready for the first opening ceremony. In the sticky heat, the most distinguished group of Australians ever seen in Western Australia stood beside the low waters of the Mundaring Weir. Arriving aboard two special trains from Perth, they gathered at the pumping house where the Bavarian Band was playing. Three federal cabinet ministers, mayors of Perth and Adelaide, General Hutton who commanded the federal defence forces, the premiers of South Australia and Western Australia, Zebina Lane, the promoter of Great Boulder, and Mephan Ferguson, the maker of thousands of the steel pipes: all were there. Mrs C.Y. O'Connor was certainly present. Nellie Melba, the singer, was in the lists of hundreds of guests, the names of whom the newspapers published in full.

THE GOLDEN MILE

Lowering a pipe into a trench dug for the goldfields' water scheme. The photo does not quite convey the atmosphere of 1903, which suggests that it depicts a later addition to the water scheme. (Photo from Battye Library, no. 10158 P)

64

The Premier of Western Australia, Walter James, had generously invited Sir John Forrest, now the minister for defence in the new federal parliament, to open the scheme. Lady Forrest turned a small wheel that started the pumping machinery, and the band played 'See, the Conquering Hero Comes'. The water flowed, the beer and wine flowed in the banquet tent, and speeches were made within the sound of the noise of the engines throbbing. The editor of the *West Australian*, Dr J. W. Hackett, said amidst laughter that he was surprised to see in the marquee some politicians and public men who used to discuss the water scheme 'with a murderous gleam in their eyes'. Sir John Forrest paid his tribute to the dead O'Connor, regretting that 'the great builder of the work was not among them that day to receive the honour that was due to him'.

Many who were present genuinely thought that the scheme now completed was the greatest undertaking in the history of Australia. The few who lived long enough to see in 1949 the beginning of the Snowy Mountains Scheme on the other side of the continent would not have altered their opinion. For Western Australia with its small population, the goldfields water scheme was bolder and more ambitious than was the Snowy Mountains Scheme for an Australia of eight or nine million people. To the engineering world of 1903 the goldfields scheme had a magic which even the much larger Snowy scheme could not provide half a century later. Often O'Connor's scheme was compared with the mighty Aswan dam just completed on the Nile in Egypt. The dams were not really comparable, but such was the length of the Western Australian pipeline that the two schemes resembled one another in boldness and magnitude.

The real opening of the water scheme could take place only on the goldfields. On the afternoon after the opening of the dam, two special trains steamed out of Perth carrying the visitors. Seventeen hours of travelling in a hot jolting train lay ahead of them. Reaching Coolgardie in the early hours of daylight on Saturday 24 January, they were feted and then arranged on seats so that they could see a street procession that included two brass bands, the men and engines of the fire brigade, camels and horses, all parading past those grand stone buildings which foretold a future that was already retreating. Sir John and Lady Forrest were placed in a decorated cart and drawn along the street by 20 children dressed in sailor suits, after which ordeal he turned on the tap at Coolgardie.

But the endurance test of eating and speech-making was not over. At 1.30 in the afternoon the trains left on the short journey to Kalgoorlie. Nearly an hour later the passengers could see the skyline of the Golden Mile, and the tall poppet heads like fire-watching towers, the smoke stacks fat and thin, and the white heaps of tailings shimmering in the sunlight. Few of the visitors from the eastern states had ever seen such a strange and memorable sight.

In the blistering heat — it was 106 degrees in the shade — women, men and small children of Kalgoorlie and

Boulder climbed the winding foot tracks up the steep slopes to the small reservoir on Mount Charlotte — the final destination of the water pumped from the coastal ranges. Along the approaches to the reservoir, as 5 o'clock came near, 12 000 people were said to be present. It was a remarkable crowd: as the mines worked on Saturday afternoon and the shops were open, so many people could not be present. The local people sensed that perhaps never before in one remote inland town in Australia had gathered such a collection of the nation's leaders, and as the leaders arrived one by one at the foot of the little round reservoir the crowd cheered, sometimes calling aloud with a shout of recognition the names of people whose faces they had seen only in newspaper sketches. Forrest was cheered, Sir William Lyne was given an ovation and, to the amusement of the reporters, that stout and witty Sydney politician, soon to be Prime Minister, George Reid, was hailed by the crowd and 'called familiarly by his Christian name'.

At the final ceremony Forrest spoke in his strong ringing voice — of course there was no microphone — and then grasped the large wheel which turned on the water. Even he, the bluff old explorer, felt moved by the scene before him, the circle of upturned brown faces, the water gushing, the round after round of cheers, and the burning sun. Perhaps he thought how only a few years ago, almost within sight of the hill, people had died from thirst.

How could they fit into the one banquet hall, on that sweltering evening, the 500 or so chairs and all the long tables needed for the visiting dignitaries and the representatives of Kalgoorlie, Boulder and the mines? High officials had gravely debated the question for days. In the end they selected the vast shed, the so-called 'car-barn' where the street trams were housed. That night the atmosphere must have been like that wonderful January day in Sydney in 1988 when the two hundredth anniversary of the British settlement of Australia was commemorated. Both elation and humility were in the air. There was also a strange episode which the headlines of some newspapers quaintly announced this way:

MUNICIPAL BANQUET AT KALGOORLIE

A REGRETTABLE INCIDENT

IMPORTANT UTTERANCE BY
SIR JOHN FORREST

HIS STRICTURES ON THE MAYOR

The mayor of Kalgoorlie, N. Keenan, taking the chair in the big iron-roofed hall, had spoken the unspeakable. And the whole nation heard about it. He pointed out, with courtesy, that the new water was not cheap and therefore would not work the expected miracles at the gold mines, and so the people of Kalgoorlie and Boulder should not be 'called upon to accept the scheme entirely with feelings

THE TORMENT OF THE WATER KING

A celebration of water: the municipal banquet at Kalgoorlie in 1903.

of gratitude'. Sir John Forrest could hardly believe his ears. He interjected with indignation. When it was his turn to speak he hinted that the water scheme was a Godlike project. He implied that it was almost blasphemy for a mayor sitting in an easy chair with iced water or champagne in his glass to speak so churlishly on behalf of all those miners 'living underground', men whose families would rejoice with Sir John in the water now gushing into the reservoir.

In one sense, Sir John was right. Miners and housewives did rejoice. Babies would have rejoiced too, if they had known the significance of the fresh water filling the reservoir on the hill overlooking the city. It was a wonderful day for the goldfields. But the mayor and a host of residents of the goldfields, while happy to celebrate that day, were trying to say, but not sure how to say it, that the scheme was not quite as marvellous as Forrest believed.

Too much had been expected of the great water scheme. It was not the triumph so eagerly anticipated. Something had gone wrong before the first heavy steel pipes were placed end to end on the hard earth and the rock.

Today it is almost a heresy to examine critically Charles O'Connor and his great work. Now he is sanctified, whereas in his last years he was almost crucified. Perhaps he was treated too harshly in his lifetime and is now treated too sympathetically. It is important now to understand the summit of his life's work because the water pipeline is regarded as such an inspired feat that there is official discussion in the 1990s of pumping water all the way from the Ord River in the far north-west to Perth, a scheme requiring a far longer pipeline than O'Connor built. One of the strongest emotions — it is an emotion and not an argument — in favour of a long Ord pipeline is that it is a replica of the O'Connor scheme and therefore likely to be thwarted prematurely by the same shortsighted arguments which, if listened to, would have thwarted O'Connor's great goldfields scheme. In fact O'Connor's brilliant scheme, when examined closely, was not necessarily an object lesson in the massive use of public funds.

The O'Connor pipeline rested on the important report he wrote in July 1896. On reading the report, one begins to feel uneasy. He was over-excited. He was too emphatic in concluding that his specific plan was the only way to provide water for the arid goldfields.

The engineer who, before reaching Perth, had spent most of his working life on the green grass of Ireland and in the moist forests of southern New Zealand, was frightened of the parched interior of Western Australia. And in fact the inland goldfields were not quite what he claimed them to be. No major engineering project in Australia was designed on the basis of such arguable evidence. Kalgoorlie has not as dry as he believed it to be. Of course O'Connor worked bravely in the face of sparse evidence, but the evidence was not as sparse as he pretended.

Was he correct in emphasising that the goldfields were incapable of supplying most of the water needed by mines and townspeople? In recommending a pipeline he had exaggerated the arguments against alternative and cheaper plans for water. He simply loaded the dice. If facts were inconvenient he — or perhaps his supporting staff — hid them.

In his report of 1896, the key report on which the water scheme was based, he explained that the rainfall at Coolgardie was too low to give any hope of filling large dams. He emphasised that for the year 1895 it received only 6.79 inches of rain. And that year, he added, 'is believed to have been an exceptionally wet year' rather than a year of typical dryness. Therefore he concluded that a miserable five inches was probably closer to the annual rainfall of Coolgardie. Now if this were true, such a low rainfall would be a very powerful argument for building a pipeline from the coast. But was it true? O'Connor's report affirmed that 1895, the year preceding his report, was 'the only complete year' in which the rainfall at Coolgardie was recorded. The thought arises, after reading O'Connor's report, that surely there were earlier records of rainfall at Coolgardie. Surely a dutiful government official would have recorded the rain at Coolgardie in several earlier years. My brief search of official papers found what O'Connor must have known: that fuller records of Coolgardie's rainfall were available. But those records did not sufficiently help O'Connor's case. Therefore he remained silent about them. In fact in 1893, a year ignored by O'Connor, Coolgardie experienced 56 days of rain in which a total of at least ten inches fell. In 1894, in contrast, the rain was a mere 3.54 inches, and that in 1895 it was — as O'Connor said — 6.79 inches. Then came another change. The records (he could have obtained them with ease) showed that the first half of 1896 was relatively wet and gave promise — a promise actually realised — that the year in which he was writing would be wetter than 1895.

Clearly O'Connor was wrong, and probably knew he was wrong, in labelling Coolgardie as hopelessly dry. Certainly he could not possibly know this fact: that the year 1895, far from being wet for Coolgardie, was drier than all but two of the next 30 years. But he should have made it clear that even on the existing records Coolgardie was not as dry as he fiercely maintained.

In making a decision for which the few facts available were all-important, O'Connor behaved strangely in not producing all the facts. He himself knew that every smidgen of evidence on rainfall was vital for his planning. He even quoted the rainfall figures for Southern Cross in the hope of buttressing his erratic argument. Now Southern Cross was also in dry terrain, and had been settled longer than Coolgardie, so its annual record of rainfall went back a little further. O'Connor reported that his argument about Coolgardie's dryness was confirmed by Southern Cross where rainfall had averaged just over five inches in the last two years.

Again O'Connor seems to have been mishandling the evidence to buttress his case. An examination of the town's rainfall records reveal that once again he did not give the government the information it was entitled to receive. The rains in Southern Cross were obviously erratic, like the rain in most of far-inland Australia, and for that reason O'Connor should have ascertained all the available rainfall rather than only the rain falling in the last two years. But he decided not to quote the rainfall of the last four years — and those figures must have been in his possession — because it could have altered his picture of the climate of the wide goldfields region.

The official rainfall figures at Southern Cross read as follows:

1892: 15.10 inches   1894: 5.12 inches
1893: 14.04 inches   1895: 5.42 inches

Moreover O'Connor probably knew, when he completed his report on the water scheme in July 1896, that the sequence of two dry years at Southern Cross was probably broken. By the end of June 1896, Southern Cross had already recorded over five inches and by the end of the year its rain had exceeded ten inches. Even the evidence available to him when he wrote his key report showed that since the first rain records were kept at Southern Cross the average rainfall was about nine inches. O'Connor preferred to re-arrange the evidence. He claimed that five inches was the average. That figure had the advantage of matching the loaded average he had contrived for Coolgardie. The conclusion is difficult to avoid: O'Connor seems to have fiddled with the evidence.

We now know that in a typical year Kalgoorlie receives as much rainfall as its rival mining field of Broken Hill. Discovered ten years earlier than Kalgoorlie, Broken Hill at its peak was slightly larger in population. Broken Hill, just as ravenous for water, eventually provided its own fresh supply without the aid of a pipeline. Not far from Broken Hill in 1892 a Melbourne company built the dam. No doubt O'Connor knew about that inexpensive dam and the role it played even before he designed the scheme for part of the goldfields in Western Australia. In his report, however, he did not mention Broken Hill. Had he mentioned it, he would thereby have weakened — but not destroyed — his own argument that the heat and evaporation around Coolgardie were strong reasons restraining the building of a local reservoir.

In conclusion, could the goldfields have dammed enough rain to supply their own needs of water in a normal year? By outback standards they did have a sufficient rainfall, indeed as much as Broken Hill. Whether the appropriate dam sites could be found on the goldfields was another question. Here O'Connor was reasonable. He was inclined to conclude that a deep dam site could not be found. But his fudged statistics, claiming a freakishly low rainfall, made it almost foolish to seek the sites for big reservoirs on the goldfields. Even if the dams were built he would have argued that rarely would enough rain fall to fill them. In his key report he mentioned his own experience in building small dams on the goldfields: 'there are many of them which have never been filled, and some of them into which hardly any water has come at all.' Of course those dams were built in the very dry years of 1894 and 1895. It was therefore not quite fair of him to say that the dams had never been filled, without mentioning that most were less than two years old.

The evidence at O'Connor's disposal, then, but not shown to the government, supported a strong effort to supply water. It did not, however, support a pipeline as large and expensive as the one he planned, and defended literally to the point of death. On the basis of the evidence in O'Connor's hands — and it is easy to be wise after the event — he probably should have built a narrower and much cheaper pipeline as well as searching for a site for a reservoir not far from Coolgardie and Kalgoorlie. That he faced a most difficult engineering problem is beyond dispute. Moreover, he had the courage to face it boldly. But was he wise to tamper with the evidence in order to advance the decision he thought appropriate?

On the question of finding underground water on the goldfields, O'Connor had also made up his mind far more strongly than the evidence permitted. The shafts being sunk on the Golden Mile were finding large quantities of salt water when O'Connor sat down to complete his report. Curiously, the report does not mention those shafts, though news of their flows of water was reported almost daily in the Perth newspapers.

These shafts were the lifeblood of Kalgoorlie and remained so year after year. O'Connor wiped them out of his calculations. He was certainly entitled to be wary about them, and entitled to ask whether they would last long. But he was unwise to dismiss them as irrelevant and to shun continuing advice from the managers on the goldfields, the very men whose mines would be the main users of his water, the men who already were trapping valuable quantities of salt water in their own shafts.

Mine water, being salty, had to be condensed at heavy expense. Such water, concluded O'Connor, was far too dear. Given the cost of firewood and equipment, he emphasised that the cost of extracting the water from the salt 'would never be less than £6 to £12 for each thousand gallons'. In fact by 1902 the standard cost of turning salt water into fresh was only £2 1/4. Moreover, the salt water itself was not only cheaper than any water that could be pumped from the coast, it was also suitable in its saline condition for many purposes in the big gold mills and treatment plants that were to be the main consumers of the pipeline water. Most of the processes at Kalgoorlie could work on salt water.

O'Connor was right to point out how dear the water in the goldfields was. But he failed to understand the needs of the mines that were the main target of the water

THE GOLDEN MILE

Coated with dust and dirt by the end of the day, the mine workers of Boulder City tried to be spick and span for the special social occasions. It was easier to be clean after the pipeline arrived. This is 1905 and the evening clearly is hot (the hall windows are open), and the tables are crowded with bottles of beer and soft drink made locally with water pumped from more than 300 miles away.

THE TORMENT OF THE WATER KING

The mines were increasingly surrounded by flat expanses and pyramid hills where the waste material was dumped. The salty water used in the treatment process was visible in the tailings dumps pictured here: the Great Boulder and part of its dump, in the first years of this century.

scheme. He failed to appreciate — it did not suit his dream — the mining company's ability to make do with salt water for so many purposes. The salt water buried beneath the hot rocky surface at Kalgoorlie was far more certain in supply and more useful than O'Connor realised.

When at last the fresh water flowed from the grand pipeline at Kalgoorlie, it was not needed in the quantities O'Connor had predicted. Such a quantity was unnecessary because numerous mining shafts were, in 1903 as in 1896, the major source of water for industrial purposes.

By then O'Connor was dead. Possibly he suspected, when he decided to end his life, that his ambitious scheme was more likely to be an engineering success than an economic success.

In Kalgoorlie, which to this day is the main consumer of fresh water piped from Mundaring weir, there was more pleasure than elation at the completion of the water scheme. Its mining leaders paid tribute to O'Connor and Forrest. Many of its citizens rejoiced in the possibility of a weekly bath, now that water was cheaper. Soon there appeared town gardens and small private gardens, green swards at the two racecourses, and rows of trees in the streets. Boulder, Kalgoorlie and Coolgardie were now more suited for wives and children. The pipeline too was a vote of the government's faith in the big gold towns, and foreign investors noted that confidence.

Perth felt proud of what it had done for the goldfields — indeed the congratulations in Perth were perhaps heartier than in Kalgoorlie. Perth's salvo of three-cheers disguised the fact that Kalgoorlie all along had been quietly solving its water problem, with little acclaim. Until 1903 its main source continued to be the salty water pumped from mines or from water shafts specially sunk near the Golden Mile. Even the visitors to the big mines could see, long before the pipeline arrived, how valuable was the salty water. In the mills they saw the snowy deposit coating the cloth of the filter press. The snow was salt. A foreman walking past a labyrinth of metal pipes in the treatment plants could tell where a pipe was leaking because a thin stalactite of salt pointed to the slow drip. Similarly the vast dumps of tailings, slimes and residues which surrounded the mines were covered with a white crust formed by salt, for the sandy waste material was deposited onto the top of the spoil dumps by a fast flow of salt water. The salt was even a help, for it formed a crust that prevented the light sands from blowing away in normal winds and shrouding the towns.

The big mines had learned how to live with salt water long before the fresh-water pipeline arrived. Nearly all waste steam was converted into fresh water. All the big mines had condenser plants on a mammoth scale — strange water-factories removing the salt so that boilers and other vessels requiring fresh water could be fed. Curiously, at the main mines the water could be bought by the public at a cheaper price than the sum about to be charged by the government's pipeline.

Along the pumps from the reservoir at Mundaring came more water than the goldfields could use. The scheme was too large, too expensive. Half of the huge capital outlay could probably have produced sufficient water to serve all the needs of the goldfields and the more remote rural towns which also tapped the pipeline. Moreover, each gallon of water was about half as dear again as the price promised by Forrest. To the disappointment of the government in Perth the sales of water were far below their expectations. The big mines along the Golden Mile gladly bought substantial quantities of fresh water pumped from the distant weir. But in the opening months they used even larger quantities of salt water from their own mines. The fact remained — their own salt water was cheaper. Admittedly it slowly corroded the machinery and the iron tanks in which water was stored but it was too cheap to discard. This was the water which O'Connor refused to admit into his calculations.

Six months after the pipeline was opened by Forrest, the mines came to an agreement with the government. They promised, for the next three years, to cease using their own salt water, allowing it to run to waste. Instead they would use the expensive fresh water from the pipeline, even in those parts of the treatment plant were the salt water itself was cheap and effective. In return for the monopoly of supplying water to the mines, the government offered the mines a daily minimum of half a million gallons for the bargain price of five shillings for each thousand gallons. The deal was signed, and three years later was renewed. It was again renewed in 1909, by which time the government decided that the losses on the scheme must be halted. The price of a thousand gallons of pipeline water was raised to seven shillings, and mines willing to pay ten shillings regained the right to use their own salt water for mining purposes. At the same time the government was determined to under-cut the salt water and offered a bargain price for those mines willing to use fresh water to sluice away the residues — a task for which salt water was eminently suited.

The pipeline was an heroic scheme. Australians justifiably took pride in it. The goldfields gained heart from it. 'Nowhere else in the world had so much water been pumped so far', proudly proclaimed a plaque unveiled on the scheme's sixtieth birthday. Its original planning, however, was hardly an ideal model for a great public work. On the evidence available when it was planned, it was too large and too costly. Thus, while giving security to the Golden Mile, its water was unnecessarily dear for those who used it.

The Kalgoorlie mines, by their unpredictable large scale of production and their rising demand for water, eventually saved the pipeline from becoming an economic failure. Other goldfields gained little. In the following years the pipeline was extended only a few miles beyond the hill where Forrest declared it open. There was water to spare but other goldfields could not afford to tap it. The pipeline soon reached the nearby gold towns of Kanowna and Bulong but their frail companies consumed little water. Even Coolgardie in its decline was a poor customer for water. Of the gold mines tapping the pipeline when it was ten years old, Coolgardie's mines bought less than 2 per cent of the water. No statistic shows so vividly how imaginative was the scheme but how risky was its original basis.

CHAPTER FIVE

ALONG THE CROWDED MILE

IN THE EARLY 1900s the dividends poured out. As one big gold mine began to fade, another took its place, replenishing the flow of profits. Never before in Australian mining had so many rich mines stood side by side. In 1907 five companies each paid out more than £200 000 for the year. The sum now means nothing until it is set in perspective. In that one year the magical sum of £200 000 in dividends was paid out by more Kalgoorlie mines than by Australian banks. And that was not Kalgoorlie's peak year.

In the previous dozen years a remarkable cluster of companies had each paid in aggregate what in those days were seen as glamorous dividends. Thus between 1895 and 1907, nine companies each paid dividends exceeding half a million pounds. Of the nine, six had each paid more than a million pounds. Leading the dividend ladder at the end of 1907 was Great Boulder with £2 644 000 and Golden Horseshoe — now the biggest mine — close behind with £2 520 000. They were followed by the dazzlingly rich Oroya Brownhill with £2 067 000 and Ivanhoe with £1 869 000. On no other mining field in Australia had six companies passed the million mark in dividends.

There was an intense pride in Kalgoorlie and the jobs and wealth it created around the nation. The wives of the managers did not all relish Kalgoorlie's remoteness but they knew they were living in one of the nation's nerve centres. Mining experts coming briefly from overseas were high in their praise of the Golden Mile. Ralph Stokes, Johannesburg's best mining editor, called at Kalgoorlie in 1906 and was inclined to think that in many ways it led his own home city, the greatest goldfield in the world. He acknowledged that 'the cream of Australian labour' and much of its engineering talent was gathered in what he called the 'richest square-mile block of auriferous ground in the world'.

The Golden Mile was a hotch-potch of buildings and smoking chimneys. So many big mines were close together that space was scarce. Great Boulder had the largest area with close to 100 acres, but even its surface was crammed in 1907. It had 17 shafts, active or disused, and each of the active shafts had its big poppet head with its winding gear to hoist the men and ore from below. Near the Main Shaft were boilers and chimney stack, winding engine, rock breaking machinery, the engine for the ball mill, and the compressors for sending air below to work the rock drills. And there was the change house where miners took off street clothes and put on their work clothes and the special house where they were issued with the candles that gave light in the underground workings. Within the sound of the throb of the compressor house were the office of the general manager and the residences of senior staff, the single men's quarters and the tiny dwellings of men who cared for the horses in the stables.

The noise of the goldfields worried newcomers. In 1906 the Golden Mile had 370 stampers at work, with 150 making a mighty din in the mill of the Golden Horseshoe, making it perhaps the noisiest workplace in Australia. On

THE GOLDEN MILE

Kalgoorlie counted its big steam boilers by the hundreds. These three boilers were being set in place in an unidentified mine in 1907.

the Golden Mile in addition another 50 or more big Krupp or Griffin mills rumbled away, crushing the rock.

Men who had not worked in a gold mine were surprised by the extent of the surface buildings as well as the maze of underground workings at the bigger mines. Here were shops for the blacksmiths, carpenters, fitters and pattern-makers. There stood the foundry and the sawmill where timber was shaped for the underground workings. Easily recognised were the big treatment plants with their crushers and furnaces, filter presses and cyanide plant. And where vacant spaces existed, dumps grew year by year — dumps of waste sands or slimes carried there by little chugging railways or horse and dray or even an aerial tram moving with a procession of buckets. Conspicuous were the tanks holding the water used in such quantities in the treatment process, and the big stacks of firewood, and all the chimneys, iron or brick, which filled the air with smoke. On windy days the dust almost jostled the smoke for air space.

Each company was a friend to its neighbour and also a foe. While the mines did not compete in selling their product — the price of gold was fixed on the world's markets — and while they rarely poached senior staff from their neighbours, they tried to outshine them in efficiency. Each mine was a law to itself, as its clocks and whistle-blasts revealed. Each mine blew a strong whistle or siren at various set times during the day — for example, to announce that it was half an hour until the 8 a.m. shift began, or that it was 4 p.m. and therefore time to begin the afternoon shift. The curiosity was that in 1902 each mine kept its own special time; and so at various hours of the day, a procession of hoots and blasts announced the time at each successive mine.

The government railways had its own timetable for the suburban trains which carried the shifts of miners to work, and it objected to the confusion caused by the chaos of company clocks and whistles which could vary from one another by at least five minutes. In April 1902, in the hope of persuading the mines to use a common time, the stationmaster at the suburban Golden Gate station offered to post one of his men by the telephone from noon until 12.15 so that engine drivers at each mine could ring up and learn the 'correct time'. The Chamber of Mines tried to co-ordinate the clocks. Whether it fully succeeded is not clear. For years it was even difficult to persuade the mines to adopt a uniform method of signalling between the engine driver on top and the men calling for the cage at various points down the shaft. A miner transferring to another mine had to learn a new code of signals — an error could be dangerous.

Each mine manager eyed his neighbour carefully, especially if there was a chance that his neighbour's lode, at depth, would change direction. Few of the lodes descended vertically into the depths of the earth: most descended at an angle. It was therefore possible for a lode to be right in the centre of the company's lease at a shallow depth but to move towards the neighbour's boundary at depth. Once it left the company's ground there was no compensation.

Great Boulder was probably the first big company to face the dreaded likelihood that its main lode would vanish into a neighbour's ground. Its general manager, Richard Hamilton, once noted that the main lode was 'going so quickly to the west' that at greater depths it went perilously close to the boundaries of two neighbouring companies, Ivanhoe and Golden Horseshoe. At a depth of 1100 feet it was expected to lie entirely across the boundary. In about 1900, Hamilton hesitated to construct a large sulphide mill for fear that, soon after its completion, the mine might run out of ore. No trace of this fear was expressed in the company's reports to shareholders. Miraculously, the main lode changed direction and at the depth of 1750 feet it was entirely within the Great Boulder's leases again.

As the shaft went even deeper, Hamilton shuddered again. The shaft revealed that the main lode was veering towards the Golden Horseshoe's ground. Indeed at a depth of 2600 feet, or half a mile, much of the lode was inside the enemy's ground, returning later. Similarly the Ivanhoe company lost part of one lode to the Great Boulder. It lost part of another to the Golden Horseshoe, which in turn lost, at depth, much of the wandering 'Great Boulder' main lode, sadly seeing it pass into the neighbouring Chaffers mine. It was not that the big lodes of the Golden Mile were exceptionally erratic. The difficulty was that huge lodes were crammed into average-sized leases. Like apples in paper bags they sometimes burst out. Investors who predicted these changes could make a fortune if their timing was correct.

The antics of the Oroya lode provided more excitement. It lay in ground belonging originally to a minor London company called Hannan's Oroya, which led an up and down life, being close to death in 1896, and again in 1901. Then a diamond drill hole, penetrating the unlikely host rock of calc schist, found a shoot or pipe of gold-bearing ore of astonishing richness. Curiously, the ore did not appear to be rich to the experienced eye. Narrow green veins or 'leaders' — the green came from vanadium — were its distinctive mark.

Soon after Hannan's Oroya found this wonderful bonanza it decided to merge, in 1902, with the neighbouring Hannan's Brownhill company. The merger made sense. The once-rich Brownhill had a fine treatment plant but only a small surviving tonnage of ore, whereas the Oroya lacked a big treatment plant but owned this wonderful shoot. Under the new name of the Oroya-Brownhill Gold Mining Company it explored the shoot.

No missing person was ever traced so assiduously or at such expense as the Oroya shoot. Like some of Kalgoorlie's wandering lodes, it too crossed boundaries. In February 1907, the Oroya's brilliant manager, the young American Deane P. Mitchell, was farewelled at a banquet where the reminiscences flowed. In a large shed decorated with 50

ALONG THE CROWDED MILE

A section of the Oroya-Brownhill mine in 1903, when it was a treasure house. Along this tramline came a procession of ore trucks loaded with the rich ore marked by distinctive green veins.

flags borrowed from the Boulder Racing Club, and filled with music from the Brown Hill banjoists, Mitchell accepted his silver tea and coffee service and his morocco-bound illuminated address and then began to reminisce about the fabulous Oroya. He recalled how, soon after the initial discovery, some of the Oroya men guessed that the golden pipe would pass into the nearby Cygnet lease. So Oroya then bought the Cygnet — for a song. It must have been one of the best bargains in mining history. The golden Cygnet, said Mitchell, went on to yield '150,000 times more money than had been paid for it'. Nevertheless, the Oroya pipe was almost worked out at the northern end when Mitchell left; and by 1910 the company's life seemed over.

According to some charts it was shaped more like a tilting flat pancake than the vertical carrot of the typical Kalgoorlie lode. Some observers described it as a buried string of sausages. It was rich rather than deep, and did not extend deeper than 1200 feet below the surface. Its length, however, was remarkable. Close to a mile long, it wandered its way into the leases of other companies, amazingly rich wherever it was mined. In all it yielded about 1 650 000 ounces of gold. More gold came from this rich ore in Kalgoorlie in ten years than came from many well-known Victorian goldfields in their first hundred years. Thus, the Oroya produced as much as the well-known Victorian goldfield of Maldon in its first hundred years. The Oroya shoot, along with the 'Duck Pond' in Lake View Consols, was the treasurehouse of the Golden Mile.

Much of the treasure did not securely find its way from the underground mine to the treatment plant in the broad daylight. The stealing of gold was on a large scale. Perhaps at its peak around 1900, it was widespread in the following ten years. Miners would take samples of the richest ore and conceal them in clothes or billycan and take them home. Occasionally they must have wrapped them in rags or bags and hidden them in the truckload of ore going up to the treatment plant where a friend collected the bag and hid it until it was time for him to go home. Many hotels appear to have been receivers of stolen gold. They paid for it in pots of beer.

The companies protested at the failure of the police to enforce the law against theft, but the police rarely found a miner in possession of gold. When they arrested a buyer of gold presumably stolen they had difficulty in convincing a magistrate, let alone a local jury, that it had been stolen. Public opinion along the Golden Mile sanctioned gold stealing. It was not theft — it was a supplement to the wages. Furthermore, a few local businessmen argued that it was preferable that some of the gold be stolen and the proceeds spent in Kalgoorlie and Boulder rather than vanishing to London as dividends. Only a brave clergyman would stand in his pulpit and denounce gold theft.

In Kalgoorlie many of those who illegally bought gold owned a little mine in the bush were they melted down the stolen ore, claiming it was the produce of their own mine. They then sold the gold to a bank. Others mixed the stolen gold with other ore and shipped it to treatment plants in Victoria. Much of the stolen gold was believed to reach Victoria where no questions would be asked. Every now and then an arrest was made and the offender fined or jailed. Thus Alfred Hall, a miner, was on his way home by tram from the Great Boulder. The police were suspicious, arrested him, and found that his billycan contained handsome specimens of gold. At his residence they found an outhouse with a little furnace and other mechanisms for treating gold ore. Pleading guilty, he was fined twenty-five pounds and costs.

A London journalist returned from Kalgoorlie in 1906 and reported that gold-stealing was rife. Detective-Sergeant Kavanagh confirmed in Kalgoorlie that it was 'carried on to an enormous extent'. In parliament the Labor Party was inclined to think the reports were an insult to the gold miners. A royal commission was set up to hear evidence. In 1907 it agreed that the trade in stolen gold was 'enormous' but could not guess at the sum. The Chamber of Mines was inclined to think that in all Western Australia the theft was about 100 000 ounces a year, and backed its own estimate with circumstantial evidence. The campaign against gold-stealing was intensified in 1907, but its success was anybody's guess. As the mines exhausted the Oroya shoot and other rich lodes or pockets of ore, the opportunities for theft were probably reduced. Occasional arrests were to show that gold-stealing was a major sideline, decade after decade. It was said in Boulder that if a miner was asked, 'What's your hobby?', he just winked.

For the companies the pockets of rich ore were a delight but the preference was for big masses of medium-grade ore. Most of the mine managers in 1900 believed that Kalgoorlie had enough of such ore to ensure a long life. To say it would live until 1920, given the speed with which ore was mined, was to be rather optimistic. Only in the Hannan's Club when the last gin or whisky was being drunk, would an expert be bold enough to maintain — and then more for the sake of arguing and keeping the night alive — that the field would survive to, say, the year 1940.

All the time the vertical shafts of the big companies went down deeper, and from the deeper levels of the shafts went out long drives and crosscuts — we would call them tunnels — to explore for gold. Many of the secret international telegrams sent from Kalgoorlie to London reported the results or lack of results. Many shareholders living in London went to the annual meeting of Kalgoorlie companies solely to hear the latest news on that crucial question: would the gold live down to great depths? A few of the larger shareholders even travelled all the way to Kalgoorlie and went down the mines, talked also to local sharebrokers, and gleaned gossip in the Palace Hotel where most stayed.

Great Boulder, leading the race to explore far below, reached a depth of 1500 feet by 1902 and was still close to gold-bearing ore. By the autumn of 1911 its two deepest shafts went down 2800 feet. So while they still had about 2000 feet to go in order to equal the most advanced shafts at the old field of Bendigo, they were among the dozen or so deepest shafts hitherto sunk in Australia. Today the depth of a shaft means nothing to the average Australian but in those days it was a subject of widespread comment. Most adults knew something about mining and hundreds of thousands of them instinctively knew what a certain depth signified. An easy way to describe a journey in a fast cage up a Great Boulder shaft in 1911 is to point out that it was the equivalent of making a journey to the top of a structure nearly three times as high as the Eiffel Tower in Paris, which is 985 feet or nearly 300 metres in height. A modern comparison is more easily made. The longest downwards journey in the lift of a skyscraper today is at the Sears Towers in Chicago. The tower is a vertical height of 1454 feet — more than 440 metres, or just over half the height or depth of each of the Great Boulder's two shafts in 1911.

The ten deepest shafts at Kalgoorlie in March 1911 were as follows:

	feet
Great Boulder (Main)	2844
Great Boulder (Edwards)	2842
Ivanhoe	2430
Associated Gold Mines	2238
Perseverance	2190
Golden Horseshoe (Main)	2050
Golden Horseshoe (No 3)	2050
Lake View	1945
Kalgurli	1900
South Kalgurli	1818

Opposite: *The more astute shareholders of Great Boulder, receiving their half-yearly reports in the post, would unfold the printed maps of the mine to see where the gold-bearing ore was coming from. These are the Great Boulder workings at the end of 1905 when the Main Shaft (one of many shown on the plan) was down more than 1900 feet. The mine was in effect a great skyscraper descending into the earth rather than rising into the sky. The vertical shafts served as the lift-wells. At every 100 feet of depth was a horizontal corridor or level, along which the ore was trucked in the dim candle-light to the shaft or lift-well.*

The blacked-out sections had already been mined, and the diagonal stripes marked those areas mined during the year 1905. In the workplaces or stopes the miners' jack-hammers driven by compressed air drilled the holes in the rock. The holes were filled with dynamite, the ore was blasted, then broken into smaller pieces, and then hand-shovelled down the ore passes to the level or corridor below, where it was pushed in trucks to the main shaft or lift-well.

With the passing years the centre of gravity of the mine moved downwards, and more ore had to be hauled from greater depth, and elaborate arrangements had to be made to improve the ventilation of the deeper places where men worked.

ALONG THE CROWDED MILE

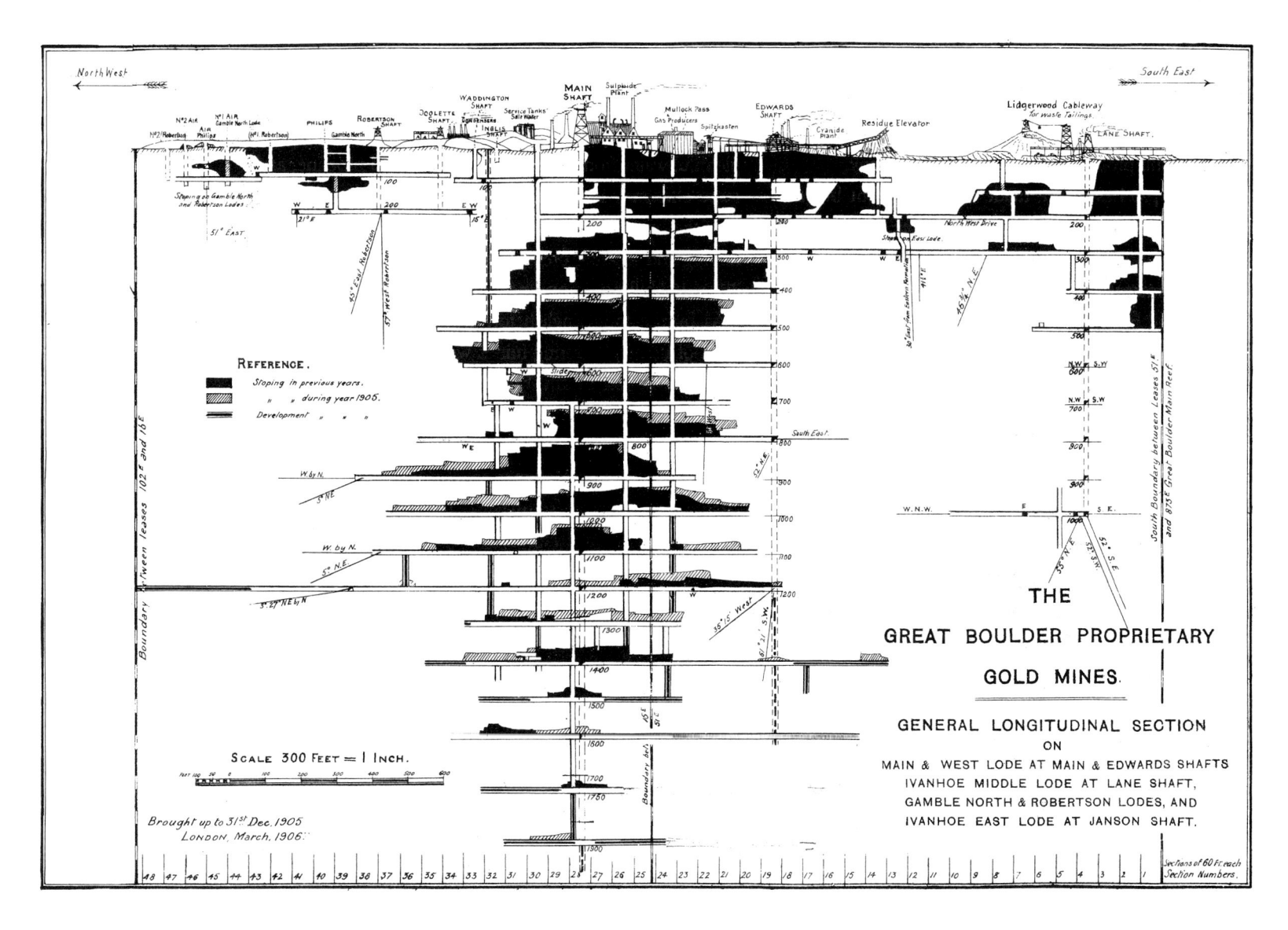

Here was an unusually deep field. Even half a mile below the surface of the earth, masses of payable ore were being found by the exploration as it extended downwards. Thus while far from as rich as at the surface, the goldfield at depth was still profitable for an efficient company.

The rich patches of ore made the excitement but the field depended on the masses of poorer ore which earlier promoters shunned as worthless. Even in 1896 a few mining commentators predicted that Kalgoorlie's future lay in mining big tonnages of low-grade ore. One of the first of these wise men was Harry C. Rhys Jones, a journalist from Ballarat. A chubby young man with a luxurious moustache twisted into wisps at each end, he predicted in Perth's *Mining Journal* that in the West the supply of ore carrying an ounce of gold to the ton was 'practically inexhaustible'. In Kalgoorlie such ore was not yet payable because of the very high wages and the dear water and fuel, but he rightly glimpsed the future. However, he did not live to see it, for the next year he died of that devastating illness of the gold camps — typhoid fever.

Of all the rock impregnated with gold, only a small percentage averaged as much as one ounce to the ton. Most ore contained well under an ounce of gold to the ton, and probably became a little poorer with depth. It could be worked with profit as long as the cost of mining and treating each ton of ore was steadily whittled down. In the early 1900s the big companies cut their costs everywhere — in buying supplies, in using better mining equipment, and in improving their ways of extracting gold in the treatment plants. Some of the dramatic cuts in the working costs came from cheaper freights on the government railways, cheaper steel and cyanide and explosives, plentiful water from O'Connor's long coastal pipeline, and more efficiency in the underground mines. But that was only the beginning. In the space of three years the cost of mining and treating ore at Great Boulder was to fall from 38 to 26 shillings a ton. Incidentally, Great Boulder measured by the long English ton of 2240 pounds, whereas most of its neighbouring mines were turning to the short or American ton of 2000 pounds, thus making comparisons a trap for the unwary.

One way of cutting costs was to increase the scale of operations. There are economies of scale, though not as large, in mines as in factories. In 1902 the biggest mines were not really on a large scale by present underground criteria. Each month they mined and treated less than 12 000 tons of ore, but by 1906 the Golden Horseshoe, a glamour mine, was treating over 22 000 tons a month.

As the scale of operations of the big mines increased, the landscape of Kalgoorlie was altered. Even in a rich mine, far less than 0.1 per cent of the ore consists of gold, and so somewhere a resting place must be found for more than 99.9 per cent of the finely-crushed rock or sand from which the gold has been extracted. Those mines that had no spare space for these tailings and waste materials pumped them to nearby waste ground and began the

ALONG THE CROWDED MILE

Dust and smoke gave a blur to this photograph taken not long after 1900. A thin veil half-hides the rising tailings dumps. In the foreground stands the poppet head of the Iron Duke shaft. In the revival of the 1930s, Iron Duke was to be one of the hopes of what became the Western Mining group.

process of hill-building there. Quickly a range of man-made hills arose. Each month the long hills of sand grew a little higher. Here perhaps was the biggest landscaping project in the history of Australia.

A miner who left Kalgoorlie in 1900 and returned for a visit 15 years later could not recognise the skyscape around the Golden Mile, such was the emergence of new hills and plateaus, barren in appearance and a source of dust when strong winds were blowing. While a few companies decided to build their tailings dumps several miles from the nearest suburb, others added to their own mountain only a stone's throw from houses. The Associated dump became notorious on windy days for its suffocating effect on the suburb of Trafalgar.

Most directors probably did not know about the dust storms their mines had helped to create. Virtually all the directors lived in the British Isles. No important mining field in Australia had hitherto been so controlled from London. The head offices of most Kalgoorlie mines were in the 'city' end of London, with a postal address of E.C. Just before the First World War, the head office of Lake View and Oroya Exploration and Ivanhoe Gold Exploration was at 1 London Wall Buildings; the Great Boulder head office, like Great Boulder No. 1, was 80 Bishopsgate; the Golden Horseshoe Estates, Great Boulder Perseverance, and South Kalgurli were at Salisbury House in London Wall; and the two Associated mines were at 20 Copthall-avenue. The head office of only one big Kalgoorlie company lay outside London. The exception was Hainault Gold Mine of Glasgow. Probably half the directors of Kalgoorlie mines had not set foot on Australian soil but the chairmen, if conscientious, tried to visit Kalgoorlie about once in every six or eight years.

The kings of Kalgoorlie were the general managers of the big mines. Normally they controlled everything except the major financial issues. Initially a typical general manager of a mine lived beside the mine, within sound of the noise of the stamp mills or Krupp mills. Later some preferred to live away from the mine, in Boulder or Kalgoorlie, where their houses were conspicuously large though never mansions. Their own salaries were high and known only to themselves. Of the leading professional men in Australia in 1905 the Golden Mile might well have provided ten of the top 300 salaries.

In the early years of large-scale mining, the general managers were drawn from many fields, near and far. Some were experienced in the United States, still one of the world's great gold producers. Thus in 1897 at least three of the managers had worked at Cripple Creek in Colorado. In addition to Henry Callahan of Lake View Consols, H. A. Judd of Lake View South and Edward Skewes of the Boulder Main Reef, Callahan's successor, T. F. Hartman, was also from Colorado. W. H. Rodda, a burly bearded Cornishman, had worked both in North America and the Rand before coming to manage the Associated Northern Blocks in 1899. The Rand and its big low-grade mines was an obvious source of managerial talent: W. R. Feldtmann, born in Glasgow, had worked in Transvaal for ten years before becoming superintendent of Hannan's Brownhill Co. Most of the top men in the mines had a strong background in gold, but there were exceptions. Thus R. S. Black, a New Zealander, was in his early career a manager of silver-lead mines on the wet west coast of Tasmania before he managed the frustrating Londonderry gold mine near Coolgardie and then the Hainault and Kalgurli gold mines.

A few of the top Kalgoorlie managers had learned their skills on Victorian goldfields. Thomas Hewitson, his eyes sunk deep in his thin face, had managed one of Victoria's biggest mines, the Port Phillip of Clunes, before coming to Kalgoorlie where in turn he managed Ivanhoe and then Associated Northern Blocks. The prince of managers, Richard Hamilton, who combined quiet authority and deep courtesy, was also a Victorian. Born at the port of Williamstown, he was educated in Bendigo, partly at the school of mines. After working in Bendigo gold mills, he became manager in his late twenties of the Honali gold mine in the well-known Indian field of Mysore, returning to manage the Peel River gold mines in NSW. After a spell as a stone-fruit orchardist in the Goulburn Valley in Victoria — the house to which he finally retired in a Perth suburb was called 'Ardmona' — he went to Arizona in 1893 to manage the Canada del Oro mine and smelters. Three years later he was recruited by Great Boulder.

Hamilton was almost a workaholic, and in his early years he began work straight after breakfast and ended his day at his desk at about 8 or 9 p.m. He had a friendly way; his big black moustache concealed a generous grin; and his face remained surprisingly boyish. Many underground men, especially those from Bendigo and Eaglehawk, were proud to work for 'Mr Hamilton'. As for their wives, they were pleased to receive a nod of recognition from Mrs Hamilton, the gracious Duchess of the Golden Mile.

A typical general manager in Kalgoorlie did not visit London as often as once in a decade. Hamilton was eight years at the mine before he was invited to London to consult with directors and attend a special meeting of shareholders at the Great Eastern Railway Hotel in January 1904, where he eagerly told them about the mine. Normally the business between chairman and general manager was conducted by telegram or weekly letter. The telegram costs of the big mines were high, even though money was saved by sending the cables in a special mining code whereby a whole sentence could be expressed in two coded words. Thus a telegram simply addressed as 'Whispering, London', would find its way to the office of the Golden Horseshoe.

Every week the managers wrote voluminous reports, laced with statistics, to the head office in London. Such time was given to preparing these reports that all kinds of other clerical and accounting tasks had to be set aside

THE GOLDEN MILE

The leading managers of Kalgoorlie and other Western Australian mines assembled in Kalgoorlie on 28 March 1911 for the annual meeting of the Chamber of Mines, in effect the employers' association of the goldfields. Richard Hamilton sits in the front row (sixth from right), wearing his tweed suit and holding his hat. Other leading managers occupy the front row.

90

each Friday and Saturday. On Monday of every week a British ocean liner punctually left Fremantle for Ceylon, the Suez Canal and London, carrying the letters hurried down on the mail train from Kalgoorlie. In 1903 there were urgent requests from Boulder shopkeepers that payday for the mines should be on Saturday, because 'Saturday night is universally the working-man's shopping night'. Such were the demands of preparing for the weekly English mail that the companies preferred to pay their army of men on the third and eighteenth day of each month — unless those dates fell on a Sunday.

At the start of the century the boards of directors of some of the biggest London companies were appointing a management firm as their consultants and even as supervising managers of their Kalgoorlie mine. The firm was Bewick, Moreing and Company of London. It set up a strong office in Kalgoorlie, and its young engineers, mostly American, managed mines or gave advice to owners.

Herbert Hoover, an American based in London, was then the Bewick, Moreing man in charge of Western Australia. The son of a blacksmith in Iowa, Hoover had studied geology at Stanford University in California before setting out in 1897 to become a junior engineer for Bewick, Moreing and Company in Coolgardie. Only 22, he appeared so juvenile for his responsible position that he grew a beard and moustache to augment his age. From his office in Coolgardie he often visited Kalgoorlie, making 18 visits in the space of two months. His firm managed the rich Hannan's Brownhill mine, at which Hoover did many tasks. He was also prominent in developing at Leonora the fine gold mine, the Sons of Gwalia, of which he became manager.

After some four years on the Western Australian goldfields and in China, Hoover became one of the four partners in his firm. More and more mines in Western Australia were entrusted to his company — for a fee and a small share of the annual profits. By 1903 Bewick, Moreing and Company was in charge of two important mines on the Golden Mile and a total of 12 Western Australian mines which together produced one quarter of the State's gold. Within a few years his firm was in charge of most of the big Kalgoorlie mines — the Great Boulder was a notable exception. However, given special responsibilities by London boards who were frightened of being defrauded by their manager in remote Kalgoorlie, the firm of Bewick, Moreing was ultimately distrusted in Kalgoorlie because its London partners were believed to be using inside knowledge to trade on the share market. In 1905 a royal commission reporting on the slump in shares in the Boulder Perseverance, a big mine now managed by Hoover's company, did not outrightly condemn Bewick, Moreing, but it did not remove the causes of public suspicion.

Hoover gave much to Kalgoorlie. He was quick to pick up the essence of a problem. He was obsessed with improving the efficiency of mines in his care. Slightly pessimistic about Kalgoorlie's future, he believed, more than later evidence justified, that its gold content was declining rapidly as the mines went deeper. To offset the declining grades of ore, even more efficiency was needed. Those who saw his square jaw and sometimes aggressive manner were correct in summing him up. He pushed aside people who stood in his way; he drilled through ground that stood in his way. In the decade 1900–10, Kalgoorlie's ability to make profits from masses of poorer ore was partly his achievement.

The mines became more and more like factories and less like casinos. Each month most companies mined about the same amount of ore as in the previous month. Fluctuations in the monthly gold output were frowned upon. Stability and regularity entered all the big mines. Likewise the town life moved in the same direction. The married men now outnumbered the single men. The cost of living fell away, and so the nominal wages did not have to be so high in order to attract newcomers to the field. Boulder and Kalgoorlie became staider towns.

Stanley Whitford, a new arrival at the start of the century, described the changed mining field. In Kalgoorlie he met a friend from his home town of Moonta and was invited to occupy the spare bed in his camp near the tailings dump of the Ivanhoe mine. Yarning with his friend, and eating his meals at one of the miners' bargain-restaurants which offered 21 meals or one week's food for just £1, he soon learned the ropes. He learned how to apply for work. Most men waited outside the mine office at noon in the hope of being selected by a foreman for work on the following day. Anyone who knew the foreman had a higher chance of being given a job.

One of his first jobs was in the Golden Horseshoe where he was given the hard task of pushing a truck of ore from the underground workings to the edge of the shaft, from which point the ore was whisked up to the surface by a powerful winding-engine. In that way a new man learned the underground ways, eventually advancing — if he was willing to learn — to the actual mining of rock. After boarding at the house of an old man from Moonta — a former wrestler with 'bow legs that resembled two moons' — Whitford moved to Mrs Stanton's boarding house. Her brother was a shiftboss at the Lake View, and through him he gained work as a trucker in that mine. Eventually he became a miner working on contract — the source of the highest wages.

The personal networks tended to mean that certain mines or certain parts of the mines had large groups of men who had something in common: they were Scots or they came from Ballarat or they were Freemasons or Cornishmen. A big gold treatment plant might come to employ two dozen men from the Queen's Church in Boulder because several of the foremen worshipped there. Whitford noted that many Italians were working in the Lake View while the Golden Horseshoe or parts of it were said to be 'a sanctuary for Roman Catholics'. Other mines recalled that the contractors working in part of one mine

One of Stan Whitford's first jobs in the Golden Horseshoe was pushing the iron-wheeled trucks of ore some distance to the plat or platform by the shaft, depicted here (1901). The truck of ore is about to go up the shaft to the crushing mill. The timbering near the shaft was whitewashed to provide more light.

might consist mainly of men who drank at a certain hotel at Boulder. In a section of another mine might be found brass-bandsmen and their supporters, for in the heyday of the brass band the town of Boulder had brass bands equal to the best in Sydney and Melbourne. What was virtually the national championship was staged at the celebrated South Street Society in Ballarat, and Boulder's AWA band won in 1904, and its rival the Boulder City Band in 1905.

Curiously, on the mines the main welfare funds in the early 1900s were run not by the new trade unions but by the mine managers. Mining was dangerous; men could be bruised or injured on their first working day, but being new to the field and having no close relatives they had no rights to medical treatment. Some belonged to friendly societies who cared for them. Most belonged to no fraternal or welfare organisation, and there was still no welfare state to compensate. Accordingly the main mines set up their shilling funds. Every employee became a member the day he began work with the company; and on his first pay day and regularly thereafter he consented to a payment of one shilling per week being deducted from his pay packet. That sum was usually less than 2 per cent of his average pay and so was affordable. It entitled every employee to free treatment in hospital and free medical attention from a doctor of his choice. The miners' and other unions, unlike those in Victoria, did not operate their own sickness and medical funds. In effect a system of compulsory insurance, Kalgoorlie's fund was probably more wide-ranging than any such scheme in the capital cities.

The safety of the mines slowly improved. For each 100 tons of ore mined, or for each 100 men employed underground, the safety record was better in 1910 than a decade earlier. Nevertheless the casualties were still high. In the eight years between 1900 and 1907, 116 men lost their lives in the mines. There was no disaster: just a run of single fatalities with occasionally an accident in which two or three men died. Fire and flood caused the big mining disasters in the continent, and fortunately Kalgoorlie suffered from neither. One disaster was avoided in 1912 when fires broke out at the 1400 feet level of the Lake View Consols, where the mine props of gimlet and the saddleback sets constructed of salmon gum caught fire, giving off fumes. About 30 men were temporarily overcome by fumes before the fires were put out.

The major hazard in the underground workings was not fully understood. The main instrument for drilling holes in the rock preparatory to firing them was a heavy mechanical rockdrill that stood on its own steel legs or supports. Virtually a jackhammer, its cutting edge filled the air in the working places with a fine dust which slowly began to affect the lungs of many men. When silica was in the rock the effect could be disastrous. Several hundred jackhammers were at work right around the clock in the biggest mines, and the machine miners were especially vulnerable to the dust created. The blasting of the holes, and the later shovelling and transporting of the broken

ALONG THE CROWDED MILE

By about 1910 many of the underground miners were breaking the gold-bearing ore in huge stopes or chambers. A strong photographer's light must have been used to turn this candle-lit cavern into such a scene of brightness. The mine is the Associated Northern Blocks. Many miners working in these orebodies were almost as white as flour millers when their eight-hour shift was over.

rock added more dust to the air. Every year, youngish machine miners were dying after short illnesses. The cause of death was usually diagnosed as pneumonia.

As the mines went deeper, the ventilation was obviously inadequate. The draughts of air which might carry away the dust were too weak. A royal commission investigating the health of miners in Western Australia in 1904 spent much time at Kalgoorlie. Through its report the dust problem became a higher priority. On the Golden Mile the big deep mines ceased to operate in complete isolation, and passageways were mined, linking the lower levels of one mine with those of the mine next door. This enabled strong currents of air to ventilate the deeper workings, so much so that the air rising up through the labyrinth of workings made the upper levels rather warmer. The air in the deep Ivanhoe mine was especially improved. That company took its task seriously, even lecturing its miners on first-aid.

There was no agreement on what caused these miners' diseases. Some union officials were inclined to point to contract mining and the pressure of work in an unhealthy atmosphere. Some miners were inclined to think in the first years of Kalgoorlie that the damaged lungs were the sign of miners who had come from Victoria where the typical gold lode, being of quartz, was a slow and silent explosive of sharp white dust. As so many of the ill miners went back to Victoria, and many others who worried about their health stayed in Western Australia but moved to the healthier life of the coast or a wheat farm, the statistical record was imperfect. Some mine managers on the Golden Mile refused to pin the blame on the dusty working atmosphere, and instead they blamed the sudden changes of temperature. In their view a special risk faced the afternoon- or night-shift men being whisked to the top of the shaft in damp clothes after working underground in high temperatures. By 1908 nearly every big mine had a large change house where miners coming up from the warmth below could take a shower and change into warm clothes before going home.

The miners' disease hardly seemed to abate: indeed it became more common. Men working in surface mills using the dry-treatment process were also vulnerable. At the Krupp ball mill and the nearby roasters they inhaled fine dust through the working day. Fortunately most mills along the Golden Mile still used a wet treatment process in which the ore at an early stage was mixed with water, thus allaying the dust.

How to control the dust was a tricky question of human relations. Respirators were offered to men working in dusty places. Some men wore them but most discarded them, saying they made breathing more difficult, especially when they were working hard. It was also thought sensible to encourage miners to spray water on the hole where their machine was drilling. The idea was tried. The spraying kept down the dust but increased the humidity. In the remoter part of the deep workings the temperature was a constant 75 or 80 degrees Fahrenheit, summer or winter. The effect, said one underground observer, was like walking into a steam bath. It was also proposed that miners with tuberculosis — it was one outcome of injured lungs — should not be allowed to work underground and so spread their infection. At first the unionists were reluctant to agree. Such a precaution, they said, would deprive the man of his profession. A dedicated Victorian doctor, J. H. L. Cumpston, conducted another royal commission into miners' health on Western Australia's goldfields in 1910 and did more than anybody to persuade those managers and miners who still needed persuading. His statistics showed that underground miners were more prone than other Australian men of their age to a variety of pulmonary diseases. It was clear that the dust was deadly and must be fought. The battle was intensified but was not a crusade until the 1920s when the health of mine workers was rapidly improved.

At that time most of the 5000 or 6000 men who worked for the mines were not members of the trade unions. Stanley Whitford first joined a union in 1905, at the age of 27. Through conversations and friendships with other miners he became what he called 'an international socialist', and eventually he won a seat in parliament in South Australia, becoming a cabinet minister. Socialists, however, were not common in Kalgoorlie. Its politics were moderate, and for that reason its local politicians won friends outside the Golden Mile.

In the parliament of Western Australia the early strength of the new Labor Party lay on the goldfields. Miners and tradesmen from the Golden Mile became prominent in Labor ranks. John Scaddan, who came west from Eaglehawk and became an engine driver on the Golden Mile, was to be Premier of Western Australia from 1911 to 1916. Philip Collier, who came to Boulder City from farmlands just north of Melbourne, was to be Premier for another nine years. Indeed in almost every year from 1910 to 1936 the Labor Party was led by the representatives from Kalgoorlie-Boulder.

The simmering dispute about how the gold should be shared between labour and capital was accompanied by another, more urgent debate. Amidst the throbbing, steaming, clattering way of life along the Golden Mile there were doubts about the long-term future of the field. The first decade of the century saw wonderful achievements, but the field could not depend entirely on good managers, skilled foremen, a fine workforce, smooth industrial relations and the most ingenious machinery. The rock itself, the amount of gold it contained, was all important.

The evidence (it is not conclusive) suggests that the ore was becoming a little poorer as it went deeper. Below 1500 feet there were few rich patches to excite the markets. Moreover the cost of working the deeper ore was higher. By 1910 a slight nervousness was voiced in Hannan's Club and the mining columns of the Australian newspapers that reported the Golden Mile in voluminous detail.

THE GOLDEN MILE

Proud members of the miners' union, with their mascot. Some miners, when dressed in their best suit, wore a watch and chain to which a small gold nugget was sometimes appended.

ALONG THE CROWDED MILE

This well-dressed crowd in Hannan Street is probably waiting to see the latest results of an election posted up on a big board as the telegrams arrive from the federal capital in Melbourne. The year is 1913, and it is possibly the federal election which Labor under Andrew Fisher narrowly lost. The seat of Kalgoorlie was now overwhelmingly Labor in sympathies; and Charles E. Frazer, who first won the seat in 1903 at the age of 23, had no opponent in this election.

Frazer, originally from Yarrawonga in rural Victoria, had been an engine driver at the Boulder gold mines and a leading unionist before winning his seat and an important place in the federal cabinet.

The two most spectacular mines — Brownhill and Lake View Consols — had exhausted their richest ore. When the high-fliers fall to earth, some gloom is inevitable, even though other solid fliers remain in the air. These two dashing companies were left with big treatment plants but little ore. Sensibly they merged with companies that owned promising low-grade mines but lacked the capacity to treat their ore. In 1910, Lake View Consols, possessing a cramped lease of a mere 24 acres, merged with Hannan's Star Consolidated, itself the result of a recent merger. The new company, with its 161 acres, adopted the catchy name of Lake View & Star. It was eventually to become one of the giants of the Golden Mile but many obstacles still lay ahead. Likewise Oroya Links with its promising mine merged with Brownhill which owned a fine mill, and a horse-drawn tramway hauled the ore across country from mine to mill. But only a few mining companies could be saved by a merger.

The Golden Mile no longer was the home of opportunities unlimited. In the four or five years before the First World War, a few companies began to think that their future lay far away. The Oroya got an option on a mine in Nicaragua. The Associated Northern picked up an old mine in Mexico. A tin mine in Cornwall and a rubber plantation in Ceylon seemed very attractive to other directors who had had almost enough of Kalgoorlie. Lake View Consols, believing that the life of its rich mine at Kalgoorlie was limited, had already acquired interests in mines in lands as remote and far apart as Siberia, Burma and Peru. It also owned a potential mine at the south end of the Broken Hill field. When it sold that block of ground in 1911 for what it thought was a favourable price, it lost possession of what was to become that field's biggest and most glamorous mine, the Zinc Corporation.

Kalgoorlie's future was far less glamorous than the one predicted some ten years earlier. But many of those who knew it well remained confident, even after the evidence gave reason for unease.

CHAPTER SIX

WAR AND THE 'FOREIGNERS'

At the Golden Mile, almost from its birth, southern Europeans were at work. Newly arrived in Australia, they marvelled at the high wages. More money could be saved by labouring at a mine on the Golden Mile in one week than could probably be saved by labouring on an Italian farm for 15 weeks. It was no wonder that more Italians heard of this paradise. If they boarded English mail steamers which called at the Adriatic port of Brindisi and sailed through the Suez Canal, they paid a lower fare than English migrants to reach the promised land. In fact many of these newcomers came from the opposite shore of the Adriatic — from the land later called Yugoslavia — but initially they were called Italians.

A few Italians acquired a small mine of their own or a profitable contract to cut firewood for the mines. Pockets of Italians and Slavs were more to be found in the outback gold towns where living expenses were high and the Australians with a wife and family were attracted only with difficulty. In 1902 the Long Reef mine at Lennonville employed almost as many 'foreigners' — to quote the simple phrase of the day — as people of British descent. The wonderful Great Fingall, far east of Geraldton, employed about 180 foreigners among its 600 employees. At that time it is doubtful whether any mine on the Golden Mile employed many 'foreigners'. In 1902, Great Boulder Main Reef was probably the home of the highest proportion of foreigners. Nonetheless the social history of the Golden Mile was influenced by these workers.

The early trade unionists were worried but not obsessed by foreign workers. In some years, however, Australians had trouble finding work on the gold mines, and so they resented the arrival of foreigners who competed for scarce jobs. On many days of 1902, not a prosperous year in Australia as a whole, more than 100 Australians could be seen waiting around the shaft of certain Kalgoorlie mines in the hope of being selected for a labouring job. The protests grew. Committees of inquiry were set up by the Commonwealth and Western Australian parliaments to investigate the influx. The 'Italians', it was widely alleged, were being brought out by agents, and the big gold mines were accused — mistakenly, it turned out — of using agents to recruit men from the Mediterranean and so lower the existing wages and conditions of work. In fact the early arrivals had sent letters home to their own village to say that this was paradise, or at least half a paradise. And lo and behold, their cousins and neighbours, brothers and brothers-in-law, set out for Fremantle.

The Golden Mile had no general policy towards southern Europeans. Some companies hired a few and several managers employed them without realising it. W. A. Pritchard, the general manager of the Lake View Consols, was asked to give evidence to an official inquiry in 1902 and did not even realise until the morning of his appearance that his mine employed a few Italians. Most managers on the Golden Mile thought well of 'foreigners' as labourers but did not like to see them working underground if their

THE GOLDEN MILE

A locomotive hauling a wood train for the Westralia Timber Firewood Co. about 1910.

English was poor. The Golden Horseshoe, employing more than 40 foreigners, wisely insisted that they should speak a little English if they were to work below. In dangerous work a common language was vital for safety. As the government's inspector of mines, stationed at Kalgoorlie, vividly argued: 'Suppose an Italian saw a stone falling and could not speak in time' to warn his workmates. Another grievance against the foreigners was that they were sometimes careless with explosives.

Out of sight of Kalgoorlie, buried away in the bushland to the east and west and south, were the camps of many foreigners. They cut firewood for the mines and timber to support the underground working. The long sticks of cut wood could be seen piled high on the little steam trains which reached Boulder along a network of lines that eventually extended more than 100 miles towards the Southern Ocean. Most heavy industries in Australia depended on coal but Kalgoorlie was probably the biggest industrial town in Australia to depend overwhelmingly on firewood. In the first three decades of mining, its boilers, water-condensers and other fuel-burning processes consumed the grand total of 15 million tons of firewood. For two tons of ore one ton of firewood was required. Even the local electricity powerhouse owned by a London company was a burner of firewood. In Kalgoorlie and district the biggest male occupation was that of miner: the second biggest was woodcutter.

The woodcutters were nomads. Once an area of scattered bush land was cut out, the wood towns were moved out bodily on the narrow-gauge tramlines to a new strip of bushland. As the suitable trees were scattered across the dry terrain, a timber area was soon depleted. An old woodcutter walking through the country in 1914 could see scores of sites where once a bakehouse, portable school or general store had stood. To see a little wood-burning locomotive, sparks flying from the funnel, bringing in a long train of firewood was a familiar sight on the fringe of the goldfield. The main woods burned were mulga and salmon gum. At the Perseverance mill the firemen always said that mulga was 'hotter'. When its Cornish boilers were being heated by mulga, smoke of blacker hue poured out the chimney.

Newcomers to the Golden Mile who did not like the idea of working underground or could not find a job in a mine went along 'the woodline', sitting on the flat top of an empty firewood wagon and alert for heavy locomotive sparks that might fall on their swag. They soon learned to cut three or more tons of wood a day. Many Irish-Australians as well as southern Europeans followed the woodline. One woodcutter was E.J. Hogan, who came from the Ballarat district in 1905 and soon became secretary of a firewood workers' union, eventually returning to Victoria where he was twice Premier. The hero of the woodline was 'the wild Irishman' Jack Carroll, who in 1916 left his job as a guard on the Kurrawang wood trains and enlisted in the Australian army. In France 14 months

THE GOLDEN MILE

Those visiting the South Kalgurli mine about 1905 saw everywhere the stacks of firewood and mine timber which formed the great input of the mine, and the sandy tailings which formed the great output of the treatment plant.

WAR AND THE 'FOREIGNERS'

A constant devourer of firewood: an engine on the surface at the Golden Horseshoe in 1903. In all mines the big wheels in the engine houses helped to lower the cages in which the miners went swiftly underground to their workplaces. The compressor house, supplying compressed air to the mechanical drills with which miners made narrow holes in the rock far below, was another heavy user of firewood.

later, in the Battle of Passchendaele, he won the Victoria Cross. After the war Carroll was to return to his old job on the woodline.

In the few years before the war, more Italians and 'Slavs' travelled along the woodlines to work for the three private companies that cut and carted wood. This was probably the main ethnic enclave of the goldfields. In 1914, as a great European war came near, the likely attitudes of foreigners working along the woodline and in the mines was anxiously talked about. They were citizens of a potential enemy. The goldfields employed several regiments of men who, depending on the course of the war, were liable to be interned. No mining field in Australia, no industrial suburb in the cities, employed such a high proportion of potential enemies.

The First World War formally began on 28 July 1914 when in Vienna the emperor of Austria-Hungary declared war on Serbia. A week later Britain declared war on Germany and then eight days later it declared war on Austria-Hungary. Australia, as an ally of Britain, was thus at war with Austria-Hungary, large numbers of whose citizens were to be found within 50 miles of the Golden Mile. Most of the Austrian citizens working at mines and wood-cutting camps were from the Adriatic Coast. In Boulder City most were called Croatians: a Croatian Hall stood in Boulder. Although from the 1890s many southern Europeans had been known erroneously as Italians, or more accurately as Slavs, by 1914 the name Dalmatian was widely used. To be a Dalmatian was an essential form of defence, for it was known that Dalmatia was a very reluctant province of the Austro-Hungarian empire. This province consisted of a string of islands in the Adriatic and a narrow strip embracing nearly all the coast of the present Yugoslavia. Nevertheless, though most Dalmatians opposed the Austrian emperor's rule over their land, they were officially citizens of the Austro-Hungarian emperor: they were therefore Australia's enemy.

In the first eight months of the war maybe 600 Dalmatians, Austrians, Germans and other enemy aliens were withdrawn from the Western Australian goldfields. They were replaced partly by British migrants who had worked in timber mills in the south-west, but, losing their job, came to Kalgoorlie's mines. There, too many of them were eventually found to be 'physically unsuited' for the heavy work. Skilled labour was scarce on the Golden Mile by the second year of the war. Output fell by about 10 per cent, mainly because the number of miners fell. Eventually some of the enemy citizens were allowed to resume work in the mines. Their labour was regarded as essential. Most were seen as loyal.

The Italians at first were considered more of a threat than the Dalmatians. At the start of the First World War, Italy was an ally both of Germany and Austria-Hungary and was expected to fight alongside them. If Italy was on the side of the enemy, then a host of enemy subjects on the goldfields, especially to the north of Kalgoorlie, could pose a problem. After nine months of neutrality, however, Italy decided to fight on the side of Britain, France and Russia. There was no 'Italian problem' until 1940.

As the war in Europe became a deadlock, and as Australian emotions towards the enemy became more heated, a further attack on enemy citizens was launched in Kalgoorlie. In part it was a wartime extension of the unions' peacetime campaign that too many foreigners were working on the Western Australian goldfields — a campaign carried on annually all the way from the Boulder municipal council to the lower house of the parliament in Perth where, even in peacetime, the lower house resolved to try to limit the foreigners working in gold mines. The war gave bite to the campaign on the Golden Mile. On 22 August 1916 the men at the Associated Gold Mines called for the dismissal of three foreign miners. The manager refused, publicly arguing that his mine was already short of labour. He also voiced the view of the Chamber of Mines that a Dalmatian was simply an unfortunate victim of international boundaries — an alien in this own land.

At the Associated, miners went on strike. In the space of days the strike spread to most of the big mines. Probably the first serious industrial dispute in the history of the Golden Mile itself, it sent waves of shock through towns accustomed to settle differences by talking. There was still talking, bubbling torrents of it at public meetings, but neither side was listening. In public a total of 14 miners, six of them from the Great Boulder, were singled out as disloyal. Eventually the mines resumed work without them.

A royal commission was set up on 9 September to determine whether the 14 'alien enemy' should be allowed to resume work and then to judge whether each of the other 124 citizens of the enemy should be allowed to continue in the mines. In the banqueting rooms of the Kalgoorlie town hall the evidence was heard by a commission consisting of three mine managers, two union leaders, and the chairman, J. Darbyshire, who was normally supervising the construction of the Trans Australian Railway which, dogged by strikes, was slowly narrowing the gap between Kalgoorlie and the eastern railway network. Of the 138 mine employees whose loyalty was investigated, all but five were Dalmatians and Croatians. In the end 33 employees, some on the basis of flimsy evidence, were expelled from the mines.

Meanwhile the output of gold slowly fell. It could not help declining. The price of gold was static but the cost of nearly everything else was increased, year after year. Profits of the richer mines fell away. Even in the initial 14 months of the war the rising costs had worried the mine managers. At Kalgoorlie the price of galvanised iron increased by 64 per cent, cement by 52 per cent, flat iron by 48 per cent, Welsh coke by 37 per cent and the crucibles used in the assay shops by 36 per cent. At the same time the price of explosives and of manilla rope went up by 25 per cent. Some of these prices jumped because the Commonwealth government increased import duties. These increased costs came just when the mines were being forced to mine poorer ore. As explained by that arch propagandist, Richard Hamilton of Great Boulder, the mines had consumed their cream in earlier years and were now on a diet of skimmed milk. Inflation, he implied, was further skimming the milk.

Several of the urgent overseas supplies dried up. Mercury or quicksilver became scarce. It was difficult, after Belgium was over-run by German soldiers, to obtain the acetate of lead used in extracting gold from the roasters. In the Krupp mills the steel balls used in the crushing of rock must have come from Germany. Now they could not be imported. Fortunately some engineers, finding the essential mild steel in the form of scrap metal at some mines, went to the forge and made their own steel balls, using a steam hammer to shape them. The government inspector of machinery proudly reported that the steel balls made in Kalgoorlie were cheaper than the imports and just as sturdy.

Every goldfield in the land felt the pinch. Many closed as their costs increased. Kalgoorlie was the safest but even its margin of profit became perilously low. Its employees declined. Dividends fell away. The big mines stopped exploring — exploration was costly — and contented themselves with mining the ore they had already blocked out in readiness. The Patterson shaft in the Ivanhoe mine, deepest in the field and down about two-thirds of a mile (3664 feet) was not extended further. The famous Ivanhoe, first mine to be floated on the Golden Mile, was really a casualty of the war and was closed down early in the peacetime.

In many of the mines, especially in shallow ground, large numbers of tributers moved in and virtually leased part of a mine from the company and kept most of the profits on ore they found. The tributer was the equivalent of the proprietor of a corner milk-bar. He was an independent miner who worked hard and took risks and occasionally made good money from veins that the big company had missed. On the Golden Mile the tributer was uncommon in 1914 and common in 1918.

The woodlines, the lifeline of Kalgoorlie's engine houses, were also disrupted during the war. As most of the big mines carried only a few days' supply of firewood and took deliveries daily, they could soon be closed by a break in the woodlines.

In 1908 a brief strike by the wood carters had closed most mines for a few days, but in January 1916 came a major stroke by the woodcutters who called on the wood companies for an extra threepence a ton. In the same week, three big mines — Perseverance, Lake View and Oroya Links — had to close because they almost ran out of firewood. More mines closed as the fires in their engine houses died away. Soon 3000 miners were idle. The typical miner even more than the woodcutter felt the loss of his fortnight's wages because he had a family to sup-

WAR AND THE 'FOREIGNERS'

Installing boilers at one of the smaller mines on the Golden Mile. Clearly visible are the doors through which the long sticks of firewood would be propelled to stoke the fires.

THE GOLDEN MILE

In wartime Kalgoorlie: A public meeting held in 1917 in Hannan Street, outside the three-storey newspaper offices. The mayor, Ben Leslie, is speaking at what is believed to have been a recruiting drive. Kalgoorlie was fiercely divided on the question of whether fit young men should be conscripted into the army and sent overseas to fight, or whether the Australian army fighting against Germany should continue to consist only of volunteers.

ALONG THE CROWDED MILE

*In wartime Kalgoorlie:
Newspapers carrying the latest news of the war are being read on the balcony of the mechanics institute at Kalgoorlie in 1917, while Jack Dwyer stands by and warns them not to look at his camera.*

port. The families of many miners had no savings: to be deprived of work was almost to starve unless friends helped. After three weeks the woodcutters won their threepence which earned them, in all, an additional £6000 a year, but the strike itself had cost six times that sum in lost wages in the mines and woodcutters' camps.

Firewood was again a worry in 1918, the last year of the war. It seemed likely then that all able-bodied Italians aged between 19 and 45 would be called up for military service for their homeland. As four of every five firewood cutters were believed to be Italian the supply of fuel to the mines was endangered. Late in October 1918 the Defence Department set up a panel on the goldfields to decide which Italians could be considered essential for the war effort. But the expected exodus of Italians from Kalgoorlie, and all the dislocation predicted, did not occur. The war ended in November 1918, with unexpected speed, and in all the celebrations in Hannan Street nearly everyone thought that the old prosperity would return. Eventually it did return, but not to the Golden Mile.

CHAPTER SEVEN — IN THE PATH OF THE WHIRLWIND

KALGOORLIE HAD BEEN one of the most prosperous towns in Australia in the 15 years before the war. During the war, with the price of gold fixed but the costs of mining that gold rising quickly, it ceased to be spoken of with envy. Wages did not keep pace with the rising cost of living. The unions looked for ways of strengthening their bargaining position for the coming battle.

In Australia the first year of peace, 1919, was not matched by industrial peace. The message of the Russian Revolution, less than two years old, was in the air. Unemployment was widespread. Real wages had not yet returned to their 1914 levels. Even Perth met serious industrial unrest in the tramways and the wharves. In Australia probably more days were lost through industrial stoppages in 1919 than in any previous year — certainly since records were systematically kept.

Kalgoorlie could not escape the unrest. The Australian Workers' Union, now the main union on the goldfields, tried to strengthen itself by a crusade against non-union labour. Whereas the Golden Mile itself employed maybe one free worker for every seven unionists, the AWU wanted the score to be seven to nil. On 6 November 1919 there was a great march of unionists, tramping from mine to mine, in the hope of clearing out the non-unionists. An attack on a returned soldier widened the dispute. Late in the day when the heads of the mines and the unions met in the office of the Chamber of Mines a crowd rushed the building, broke the fence and many windows, and surged into the conference hall. The Chamber of Mines in its monthly report denounced a 'howling mob' and the 'day disgraceful in the annals of the fields' as INDUSTRIAL ANARCHY ON THE GOLDEN MILE.

For several days the local police were outnumbered. About 100 additional police armed with rifles were rushed by train from Perth, and from the local people a few hundred special constables were enlisted. It was the first time such industrial strife had ever been experienced on the Golden Mile. Calm was imposed, the streets became orderly again, but the mines were closed for about nine weeks. A year earlier such turbulent events, well known on some mining fields in eastern Australia, could not be imagined in Kalgoorlie.

Meanwhile a miracle was happening. For more than a century the price of an ounce of fine gold had been generally fixed throughout the world at 85 shillings (or 84 shillings and eleven and a half pence, to be exact). Early in 1919 the gold market was shaken by the unpegging of the American dollar. The English pound, always worth about \$US4.76, moved to \$4.35 in the space of five months and continued to move. Before long the English pound and its close ally, the Australian pound, were devalued against the American dollar by more than 25 per cent. In effect an ounce of gold was now worth much more in New York. Permitted by the Australian government in February 1919 to sell their gold to New York or Shanghai or wherever the return was largest, the Kalgoorlie and other Australian gold mines snapped up what they called 'the premium on gold'.

In the first 16 months of this free gold market, Kalgoorlie mines earned nearly 109 Australian shillings instead of 85 shillings for each ounce. Thus 1920 was a prosperous year for Kalgoorlie. While the high price of gold did not equal the recent rise in the cost of producing that gold, it was valuable compensation.

The employees whose standard of living had fallen since 1914 now had the chance to catch up. In 1920 they were awarded higher wages, as well as an overtime rate applying especially on Sunday when much of the maintenance on the mines was done. They also received the privilege, not yet commonplace, of an annual holiday of two weeks on full pay. These concessions did much to restore harmony.

So the Golden Mile had a breathing space. But soon it was puffing again. The costs of mining soared, partly because of the higher pay. In Kalgoorlie and Boulder 600 men lost their jobs in the year 1921. In the twin towns some 400 houses were dismantled and removed closer to the coast. To increase the perils of the surviving mines the gold premium — the difference between life and death — slipped back to ten shillings an ounce. However, the costs of mining and the wages did not fall.

In some mines that were once spick and span and orderly, the slapdash was now seen. In 1921 in the show mine, the Great Boulder, the underground manager, John Warrick, reported that absenteeism was high. He noted that on a recent pay day 266 men worked underground, but on the next day — with pay presumably being splashed about in the hotels that were still in plague proportions — only 210 arrived for work. After the miners were whisked by the cageload up the shaft, grimy after their hard work, they walked a short way to a relatively new change room with hot and cold running water. There were 14 showers, but from 11 the metal spray had been stolen. Yet along with the slapdash was a touch of benevolence. Warrick proudly reported that one of his underground men, bearing the Cornish surname of Hocking, was still working underground in a light job at the age of 84.

In 1922 the Arbitration Court, in order to keep mines alive, reduced the minimum wages on the Golden Mile from 16 shillings to 15 shillings a day. The fall was too much to satisfy the labourers whose pay was the lowest and not enough to help the struggling mines. Sutherland of the Golden Horseshoe summed up the dilemma when he pointed out that, with the help of the premium, the gold price was about 10 per cent higher than at the outbreak of war, while the costs of mining each ton of ore were up about 50 per cent — up from 33 shillings and one pence a ton to 49 shillings and ten pence a ton. His famous mine for the next few years tottered on the brink of closure. Nothing could save it.

In 1922 a few companies closed all or part of their mine. Lake View & Star put off 300 men while it tried to reorganise. As many of these men had no savings and as the dole was not yet invented, they had to queue up to obtain food rations for their family. Other companies decided that the Golden Mile was becoming the Scrap Iron Mile, full of rusting and discarded machinery. It was almost time for them to leave. The Associated Gold Mines of W.A., once a big mine, looked to Ontario. The Ivanhoe Gold Corporation broke an unusual habit formed back in 1901. For the first time it ceased to pay its three-monthly dividend. At least the Ivanhoe had the satisfaction, after bravely following the advice of the geologist Dr MacLaren, of sinking its shaft deeper through a barren area of calc schist and picking up the main orebody which earlier had vanished from sight. There was only one hitch. In the dismal economics of the 1920s the new ore might not be worth mining. The directors in London saw no future for their once-great mine and its magnificent wide shaft about 7/10ths of a mile in depth. They sold out to the Lake View & Star.

Only one mine was safe, the Great Boulder. The annual meeting of shareholders in London in 1922 heard their long-standing leader, Sir George Doolette, make a confession which, cabled to Australia by news reporters, shocked most shopkeepers in Hannan Street. 'Lately,' he said, 'a strong movement was on foot to shut down the whole of the mines.' The plan — either devised by managers in Kalgoorlie or by directors in London — would have closed every mine, conserving the ore until what Doolette called the 'cruel time' was over. That plan, if carried out, would have reduced the population of Kalgoorlie-Boulder from more than 20 000 to 5000 in the space of two years.

The nearby ghost town of Kanowna was a warning. The state government now wished to tear up the railway line from Kalgoorlie to Kanowna because grass — in the few weeks of the year when grass was sprouting — could be seen in the middle of the railway platform and even on the main streets which 20 years ago were jammed with shoppers on Saturday nights.

Many of the older miners were tempted to leave the Golden Mile, not only because wages were no longer the big attraction of the field but because the fear of miners' disease was rising. Before the war the causes of the disease had largely been diagnosed, and initial precautions taken. In 1914 the Chamber of Mines agreed to set up a Mine Workers' Relief Fund, financing it jointly with the government of Western Australia. The funds, however, were not sufficient to finance adequately the miners who decided to abandon their occupation because of their weakened lungs. Moreover, gold mines closed or became poorer, reducing the source of new funds. Soon the scheme was useful only in Kalgoorlie where the companies and all the men contributed.

On the Golden Mile new employees had to pass a medical examination before they could begin work. The idea of an examination for all employees, old and new, was supported by the companies but refused by the unions. The idea of being made redundant did not appeal to most men in the difficult years after the war's end. In August 1922 parliament intervened. John Scaddan, the former Kalgoorlie engine driver who fell out with the Labor Party he had once led, introduced the Miners' Phthisis Act. He was now minister for mines in a non-Labor government and representing Albany instead of his old goldfields seat of Ivanhoe. One of his brothers, it is said, had suffered from miners' disease. Scaddan also knew the details of the South African attempts to fight the spread of lung diseases in mines.

Scaddan's bill called for the compulsory medical testing of all men working in or about a mine. A man found suffering from tuberculosis could no longer work in a mine. A controversial law, with deep implications for men who had to give up their work and for the companies that were battling to survive with a less experienced workforce, it was widely debated. The Act was passed but not put into effect until September 1925. Six months later a revised Workers' Compensation Act offered some financial incentive to miners with silicosis — the latest broad-ranging name for miners' diseases. They were encouraged to take up other work.

One cause of the decline of the Golden Mile did not enter the minds of certain general managers. The cause was simple: they themselves were partly to blame. One of the most adventurous mining fields the world had seen was becoming one of the most staid. The spirit had been squeezed out of most of the leaders of Kalgoorlie mining.

That painful fact was announced by Kingsley Thomas, who came as a royal commissioner in 1925 to report on the field and its future.

Thomas decided that the seven main surviving mines were like hermits, living in isolation from the outside world and not even co-operating readily with each other. There were seven London offices, seven London boards of directors, seven general managers earning high salaries in Kalgoorlie, a most expensive network of offices and administrative costs, at least seven separate mining operations almost side by side, indeed at least seven of everything when in fact one or two would be more efficient and cheaper. The generation of power using a mountain of firewood each year was especially inefficient in his eyes. The mining field had 23 different steam powerhouses, whereas he thought one efficient powerhouse would be enough. He counted 21 major shafts still in use — the Golden Mile was like a skyscraper with lifts competing with one another so close together that it was almost as if a manufacturer of lifts had designed the building to maximise his sales. Of course there were valid historical reasons for the duplication of everything on the Golden Mile but those reasons had vanished. He called for one big company, a couple of well-equipped shafts and treatment plants. Whereas the companies had met the problem of rising costs by mining only the richer patches of ore, his solution was the opposite. He said they should become as efficient as possible, mine on the large scale, and cut the cost of mining and treating each ton of ore so that the unpayable ore became profitable. In short, they should return to the solution they had adopted with success in the early 1900s.

His report was like a cloudburst at a garden party. The seven general managers, shocked, ran for shelter. After a short search they gleefully found mistakes in his report. He had not fully inspected some of the mines and so his report was inaccurate. Correctly they pointed out that the efficiency of the Golden Mile miner was still much higher than that of the black miner on the Rand — the field which he praised so highly. Richard Hamilton, dismissing the report as 'not very helpful', defended the use of horses and drays for carting ore and defended almost everything else. But most of Thomas's criticisms were probably valid. Two years later the Broken Hill mine manager W. E. Wainwright, headed a committee that signed a similar report to the Commonwealth government, prompting the Prime Minister, S. M. Bruce, to agree that it was 'a damning indictment' of those who ran the field.

A reluctance to try new ideas was now a hallmark of the Golden Mile, not least in metallurgy. Kalgoorlie had been ingenious in solving the early problem of how to extract the maximum gold from unusual rock, but its up-to-date methods of 1905 were no longer so smart in 1925. A new process, largely developed at Broken Hill before the First World War, was revolutionising the extraction of a wide variety of minerals on mining fields in every continent. It had something to offer Kalgoorlie, but there it was not tried. The field was struggling to survive. The scrapping of old treatment plants and the expensive building of new was not on its agenda. Nor was change even toyed with as the arteries hardened.

At the Kalgoorlie School of Mines in June 1923 the first experiments were conducted by A. S. Winter and B. H. Moore. The essence of the process was to crush the gold-bearing ore, agitate the mixture of sand-like minerals and appropriate chemicals in a tank or trough, make a myriad of bubbles rise, and the selected mineral grains would rise with the bubbles, leaving the residue behind. It was called the flotation process, and before long it was showing its merits on ore from the South Kalgurli mine. To the delight of the investigators the use of salt water — of which Kalgoorlie had plenty — was more effective than fresh water. The additive or agent that they found most effective in the salt water was six parts of coal tar to one of eucalyptus. Early tests showed the pleasing recovery of 94.4 per cent of the gold from South Kalgurli ore, but that was not much above the normal rate. The first company on the Golden Mile to start its own little flotation plant was the Oroya Links. The investigators from the school of mines and the Oroya Links worked together in 1925, learning by trial and error. By the end of the year the small team was almost on top of some of the pitfalls, though the white-coated men were more confident than the 'practical' ones who ran the mills.

The flotation process was being used in scores of mining fields from Scandinavia to South America. In Kalgoorlie it was still an experiment. A strange isolationism now pervaded the field. The companies did not send their senior staff to visit other Australian fields — even after the completion of the transcontinental railway in 1917 cut days off a trip to Adelaide and Sydney. When staff did travel to the big technical congresses organised by the Australasian Institute of Mining and Metallurgy they paid their own fares.

Even visits to neighbouring mines in Kalgoorlie were not common. 'I do not know what they are doing on the other mines,' said Charles Blackett to Commissioner Thomas in 1925. Blackett, metallurgist on the Golden Horseshoe, was a go-ahead man of middle age who, several times before the war, had travelled all the way to England to give evidence in mining litigation for his company; but each time he had been refused permission by the London directors to visit American mines on his way home. London's directors were as much to blame as their general managers on the field for the closed eyes. Thomas largely exempted London from his blame but the final decisions on any amalgamation of mines could be made only in London. No such decisions were made by London directors, except as a last resort.

Kingsley Thomas had his successes. He urged a stronger attack on the dust in mines and the havoc it created. His call for the adoption of the flotation process was an indirect crusade against dust. Being a wet process, it allayed the dust in the treatment plants. In the surviving mines the crusade against dust was promoted by the inspectors of mines. Ventilation was improved with the addition of big mechanical blowers. Lighter rock drills of new design sprayed water into the hole while they drilled swiftly into the rock, thus producing a trickle of slurry instead of the damaging dust. Each year, hundreds of counts of the dust were made in the mines and the mills in order to find danger spots and seek remedies. In the late 1920s large supplies of ice were being sent down to the deeper deadworkings in order to cool the working temperature. Cooler air meant less dust.

With the incessant discussion of the dust, and with the sight of miners dying before their time, young goldfields men were not so tempted to follow their father's occupation. Normally the leaders of the Australian Workers' Union on the Golden Mile were sensible and eminently practical in assessing this matter. They reported in 1925 that young men would take on trucking in the underground workings but were wary of machine-mining, which was the dustier job. Moreover the wages — once the best in Australia — were no longer a real reward for such arduous and hazardous work. So the average age of the miner increased.

Despite the hazards in the deep, dusty mines and despite the falling pay, Kalgoorlie had one remarkable asset. Its industrial relations normally were smooth. Compared to the silver-lead-zinc field of Broken Hill, and the black coal mines of coastal New South Wales, and compared to many of the copperfields which were home to the firebrands of the Industrial Workers of the World during the First World War, the Golden Mile was almost honey. Managers and unions talked freely. The strike was a rarity. A communist in Kalgoorlie was as rare as a very wet day.

Almost throughout Kalgoorlie's history the forces making for industrial peace outweighed those that stirred the pot. These influences were rarely discussed, but looking back, some of them can be glimpsed. The rich profits did not go entirely to the absentee owners. Individual enterprises flourished along the Golden Mile, and a considerable section of employees did not see themselves as solely wage-earners whose interests always had to conflict with the owners. Many miners shared in the profits through gold stealing. The miners with the most skill earned big pay because they had the incentive of the contract system. At certain periods, large numbers of tributers took over part of the mines like share-farmers. In short, a big group of miners at Kalgoorlie did not see themselves as unmistakably on the opposite side to the absentee mine-owners. Moreover, they understood the economics of gold mining and therefore did not make demands which would help to close the mines and cut off their own livelihood.

Several other reasons can be found for the smoother relations in the workplace at Kalgoorlie. For a surprisingly long period, unionism was not compulsory. It could be that the miners found they could achieve more through politics than through unionism. The goldfields after about 1901, when they began to support Labor candidates, wielded an influence on their state parliament such as no other Australian mining fields had exercised on their parliament — except perhaps in Victoria in the 1860s. In Western Australia the Labor Party was for the goldfields what the Country Party, say in Victoria, was for the small farmers. Kalgoorlie mines provided, in Scaddan and Collier, two of the longest-standing Premiers in the State's history. Much of the legislation that altered the life and working conditions of miners on the Golden Mile in the years 1910 to 1940 came from parliament in Perth and not from unions' agitations in the mines.

The smooth work-relations that helped to keep the mines alive in the hazardous 1920s were also aided by the spirit pervading Western Australia. It was still seen, rightly or wrongly, as a land of opportunity for all. The sense of grievance was generally weaker in Western Australia. Kalgoorlie reflected that spirit. Significantly, in the period from 1900 to 1930, many Kalgoorlie miners and surface hands saved enough money to move away and pay a deposit on a wheat farm. They did not feel that they were doomed to be wage-earners. They did not see militant action, through unionism, as their only hope. In attitudes to the world they were a far cry from Broken Hill and yet in the 1920s they could have mustered as many legitimate grievances as the unions of Broken Hill.

THE GOLDEN MILE

A slow revolution was taking place down below, and by the late 1920s it was becoming visible.
The dust in the mines under old mining methods can be guessed from the clothing in this scene from the Hainault mine in 1904. Two of the miners in the background have tied bowyangs in their trouser legs, below the knee, to help curb the rising dust: the nearer men wearing collars and ties are officials and are in the workings only temporarily. It is rare to find a photo of dusty working conditions simply because the photographer, seeking as much clarity as possible, waited until long after the drilling and shovelling had ceased, and the dust had settled, before he set up his camera.

IN THE PATH OF THE WHIRLWIND

A later version of the rock drill which helped to allay the dust by spraying water into the hole where the drill was churning into the rock. The price of the new, healthier technique was higher humidity and discomfort in the underground workings. In this scene far underground in the Great Boulder about 1960, water is trapped on the floor; but in old mining photographs the dampness is rarely to be seen.

The later miners, instead of carrying candles, strap onto their belt a strong battery which gives power to the lamps fitted to the front of their steel hats. In the earlier photo the miners wore soft hats which offered scant protection if rock happened to fall.

There was a fall not only in the number of mines and miners on the Golden Mile in the 1920s: likewise the big mines were no longer so big. In 1925 three mines each sent up the shafts just over 100 000 tons of ore for the year — whereas just before the war they had been twice as productive. Moreover the gold content in the ore was poorer.

The dividends fell. Great Boulder had paid a dividend of 150 per cent in 1917 but five years later it was down to 50 per cent. Such a dividend was still remarkable but not for those old shareholders who had come to expect much more. They could recall, from 1896, a period of nearly a quarter of a century when not once did the dividend fall below 100 per cent, and so in 1924 they must have read with dismay the reports in the financial press explaining that no dividend was to be paid that year. For five years in succession no dividend arrived in the mail. Such humble profits as the company earned were set aside for a rainy day. The rainy day, of course, was already there. Richard Hamilton himself walked out into the rain in 1927, retiring as general manager but remaining chairman of the powerful Chamber of Mines based in Kalgoorlie. Living on until he was almost 88, Hamilton was to see Kalgoorlie rise again.

The Golden Horseshoe Estates, to use its noble name in full, was even worse off in the 1920s. 'The Shoe', as it was called locally, had been the star of the field. It led the dividends in many years. For the decade to 1916 it mined more ore than any other mine, averaging 250 000 tons a year. In earlier years its profits made the richest mines in Victoria seem dowdy. In a brief slump it paid no dividend in 1911, 1912 and 1914 before resuming its profitable life. Then in 1925 it made a loss. It had tried to mine the richest ore but even that was not good enough. It was left with big blocks of ore averaging a mere eight pennyweight, and if they could be mined only at a loss then 450 men would lose their jobs. The general manager, John Sutherland, the quiet bachelor who had once been the inventor-hero of the field, tried hard to find ways of retaining his workforce. The government in Perth was asked to guarantee a bank loan for £25 000. The sum now seemed puny, equalling less than one month's dividends in the palmy days of 'The Shoe'.

Nothing could save the Golden Horseshoe. Its ore was not rich enough to withstand the rising costs of mining. Dearer wages and supplies had helped to increase costs of each ton of ore by 85 per cent in the last ten years. Alex Montgomery, the State mining engineer, inspecting the doomed mine just before Christmas 1925, estimated that half of that financial burden actually arose from the fact that the company was handling a lower tonnage of ore, which in turn meant that the cost for each ton was higher. 'The Shoe', forced to mine selectively, thereby increased its costs for each ton.

Hundreds of tributers were now at work, scavenging ore that the big companies had missed or working little mines

IN THE PATH OF THE WHIRLWIND

In 1927, beneath the poppet head of one of the Great Boulder shafts, white-haired Richard Hamilton, about to retire, sat down in the second row, surrounded formally by the senior men of his mine: they included officials from the general office, the mill, the underground, the assay laboratory and the engineering shops.

A few others were invited to join the scene; and in the front row is the resident dog and the office boy wearing short trousers and holding a man's hat; and standing at the side and back are a few of the operatives who received weekly wages rather than an annual salary. On most mining fields in the 1920s the 'workers' would not have been invited to join the picture. Here was another quiet sign that divisions between bosses and men on Kalgoorlie were not sharp.

123

on the edge of the field. Like mice gnawing the edge of the cheese they often worked on narrow margins of profit, nibbling away for an income which proved to be lower than they would have earned as contract miners. Occasionally they found a rich little lode and made a small fortune in the space of the year — enough at least to pay off their house and take the family to Perth or even the eastern states for a holiday. Some tributers lost most of their gold income through the sums they paid the mining company in royalties and in charges for the use of the steam to drive the engine and supply the compressed air in the underground workings. Albert Duke, who claimed in 1921 to be the oldest tributer 'on this belt' and was working in a remote corner of the Lake View & Star, used strong words about the way the mining companies treated his fellow-tributers. He called them 'Shylocks'.

The tributers multiplied. In 1925 two parties of tributers were working in the Paringa mine, two in the Brownhill, and five in the North Kalgurli. Each morning many tributers were to be seen riding a bike or walking to the old Iron Duke, the Oroya Northern Blocks or other leases. On the famous Perseverance — usually called 'the Percy' — 40 per cent of the ore was mined by the tributers. Almost everywhere these small groups, earning in a bad month absolutely nothing, were hard at work. By 1927 the tributers earned a vital part of the field's income. Like share-farmers they were taking the risk that the big mines no longer took. Without them much of the golden 'farm' would have been idle.

Kingsley Thomas had called for fewer mines. In one sense there now were more mines, for most of the big mines were effectively subdivided into many small mines worked by teams of tributers. These teams gouged out the richer ore, following the little veins of gold with enthusiasm and shunning the larger areas of low-grade ore which Thomas vainly hoped would revive the field.

In Australia the gold industry had almost collapsed. The Kalgoorlie field, no matter how ailing it seemed, was the healthiest of all. It produced four of every five ounces of Western Australian gold at the end of the 1920s. But compared to the four years 1901--04, it was in darkness. In those years gold had provided more than 80 per cent of Western Australia's export revenue but now it was closer to 1 per cent.

Kalgoorlie was beginning to live in the past. In some weeks in the Perth or Melbourne newspapers the goldfield was more the topic of reminiscences than of reports on current mining operations. Paddy Hannan's death in the Melbourne suburb of Brunswick in 1925 inspired reminiscences all around Australia. Still a folk hero, the man who had rescued tens of thousands of people from the grip of a depression, he belonged to an era of discovery which, most people said, would not come again. They pointed out with nostalgia that in the one-third of a century since he found Kalgoorlie no major goldfield had been found in Australia. Another stream of reminiscing ran fast in the

In the 1920s Kalgoorlie might be 'tattered at the corners' but many of its activities were conducted with the old sense of style. Here is a dinner in the Palace Hotel, itself seeming slightly out of date with its pressed-metal ceilings and its line of ceiling fans standing out like aeroplane propellers. But the tables are decorated with specially-printed menus and bowls of flowers and fruit, the five neat waitresses are ready to spring to attention, and nearly every man is wearing a black tie. Among the diners only one woman can be seen.

next couple of years when the two best known of the managers — Mr Hamilton and Mr Sutherland — moved their residences to Perth. To be called Mister with respect was the equivalent of a knighthood on the Golden Mile.

The appeals to the past, the recounting of anecdotes and the handing around of knick-knacks and faded photographs, became a deluge in 1928 when Herbert Hoover, well known on the Golden Mile and at the northern town of Leonora, was elected President of the United States. Some memories of him gained in the telling. A poem he is said to have written to a Hannan's barmaid was discovered — or fabricated — and quaintly described as exotic and erotic. It was about as erotic as a newly-opened sardine tin but was published, the entire 34 lines, in Arthur Reid's fine book of nostalgia, *Those Were The Days*. The last verse began with the alleged confession by Hoover.

> *It was but a summer madness,*
> *that possessed me, men will hold,*
> *And the yellow moon bewitched me,*
> *with its wizardry of gold.*

Alas, gold seemed to have lost its wizardry. The Golden Mile clamoured for another discovery or a revolution in the economics of gold mining. Neither event seemed very likely when Hoover entered the White House.

The twin towns of Kalgoorlie and Boulder were now tattered at the corners. Many buildings on the outskirts of the town no longer had tenants as the population dwindled. Some houses were not being repaired, halls and churches needed reroofing, painted signs on some shops were peeling, and here and there a shop was empty. Loyalty to the goldfields was high, but faith in its long-term future was lower than ever before.

Towns in need of repair were vulnerable to the tropical cyclones that, every decade or so, strayed down from the far north-west. Early in the afternoon of 10 February 1928, Boulder City, Kalgoorlie and Coolgardie were hit by the most severe cyclone in memory. The storm could be seen approaching with its warning cloud of dust. The cloud became a wall. In three minutes of fury the gale lifted scores of roofs, blew down the wooden walls of scores of houses and shops, and flattened furlongs of fences. Not one sheet of iron was left standing on the fence surrounding the Foundry cricket ground. Some streets were in shambles. The newspapers for days reported the oddities: the wind took away Mrs Williamson's house in Hare Street but left her just one room, and it tore much of the roof from the railway station and wrapped it around the trans-Australian train about to depart. The cyclone tossed outhouses into the air and pulled down electric wires, leaving Boulder without electricity after dark. A racehorse broke its neck, and at the magnificent racecourse, the 'tote tower' was smashed. The debris injured people. Two churches in Coolgardie and one in Kalgoorlie were toppled. Why nobody was killed by the flying sheets of sharp-edged roofing iron was a mystery to many bystanders.

Men must have been close to death when the wind toppled three of the large chimneys at the Boulder Perseverance and suddenly cut off electric power from the engine rooms — vital to the mine. At the Perseverance the men working underground had to show that very quality embodied in the name of their mine. Since the cages in the shaft were deprived of all power they had to climb a long distance, hand over hand, to reach the surface.

The economic cyclone came in the following year — the start of the world depression. Herbert Hoover, sitting in the White House in Washington, seemed powerless to control the world's economy that had seemed so promising on his inauguration day. For Kalgoorlie this was a cyclone in reverse. Magically it nailed roofs back on buildings, erected hundreds of new houses on vacant lots, filled shops with goods, and erected tall poppet heads over long-deserted mines.

THE DECADE WAS UPSIDE DOWN CHAPTER EIGHT

IN 1930 ONLY ONE big town in Australia had reason to think that the oncoming depression would bring it more good news than bad. That town was Kalgoorlie. Its more astute residents knew that a new depression would cut the costs of mining gold, thus reversing the trend that had slowly crippled the Golden Mile. Moreover they were also lobbying the Federal government to raise the price of gold on the Australian market on the grounds that a federal subsidy for gold would aid the troubled Australian economy as well as their own mines.

In Canberra in December 1930 the Scullin Labor government finally agreed to subsidise the price of gold. Its Gold Bounty Bill promised to pay a subsidy of one pound an ounce, or close to a bonus of 25 per cent. Though it was to be paid only to those mines that produced more gold than in the three years 1928–30, and paid only on the additional gold they produced, it was a windfall for Kalgoorlie. No federal act of parliament had ever brought such heart to the Golden Mile.

Early in 1931 a bigger bonus reached Kalgoorlie, finding its way to every shop counter. The Australian government was forced to devalue the Australian pound against the English pound and nearly all other foreign currencies by about 25 per cent. The first major devaluation in Australia's history, it earned for all exports, including gold, an additional 25 per cent on the overseas markets. The price of gold was so enhanced that the promised gold bounty of one pound an ounce, conditionally awarded by the federal parliament, became less necessary. The rate of the bounty was halved in July 1931 and suspended in September 1932. As the depression went on, the price of gold continued to rise, doubling between 1930 and 1934. It was Australia's only commodity to rise in the course of a depression which saw a fall in the price of nearly everything else from wheat to soap. As a result gold mining was potentially as profitable as it had been before the First World War.

Kalgoorlie blossomed. The mining companies that had clung to the field made profits again, and some tributers made small fortunes. In Boulder the price of houses and hotels crept upwards and then jumped. New corrugated iron for roofs and fences, and paint too, was sold in massive quantities. In crowded shops on Christmas Eve, big quantities of toys were sold. Record crowds attended the illegal games of two-up that flourished at the Half Way and other secluded spots near the Golden Mile. According to the historians of two-up at Kalgoorlie, the game drew profits from gold stealing — highly profitable again with gold so dear. Miners who earned only £5 a week could be seen betting with bank notes of £10 or £20, and occasionally taking home a winning sum equal, in 1985 values, to about $14 000. People who came to Kalgoorlie to try to escape the depression were astonished to find that they were escaping into a financial boom on the goldfields.

At weekends the townsmen who motored along bush roads within a radius of 50 miles of Kalgoorlie could see, after dark, the lights of prospectors and miners camped at hundreds of points in the scrub — men looking for new lodes or reopening old mines. On 16 January 1931 two of those prospectors arrived in Kalgoorlie in a car, accompanied by a Coolgardie constable. When the car pulled up at the post office, people crowded around the running board and craned their necks to see the cargo so carefully guarded — the largest nugget ever found in Western Australia. Called the Golden Eagle, it was too cumbersome to fit on the gold scales at one bank and had to be taken to another bank for formal weighing — it weighed more than 70 pounds.

In each sharp turn of the economic wheel, new or neglected gold mines are likely to come to the fore. The latest champion of the Golden Mile was the Lake View & Star. The offspring of a merger of 1912, it had battled along under the alert management of the elderly H. E. Vail, one of the two managers who escaped the quiet scorn of the royal commissioner Kingsley Thomas. Vail had bright ideas but little money to implement them. Fortunately the long-term prospects of his mine impressed one of the world's largest gold companies, Consolidated Gold Fields of South Africa. In 1926 this London-based company bought a large interest in Lake View, to use its short, popular name, and also bought into the promising Wiluna mine far to the north. Few people predicted that these two were to become Western Australia's show mines a decade later.

It was John Agnew who brought this South African expertise and profits to the aid of Lake View and his old field. A New Zealander, a graduate of the fine school of mines at Dunedin, he had come to the goldfields in about 1900 and worked for Bewick, Moreing when they were the management kings. Still in his twenties, Agnew became an acolyte of Herbert Hoover, following him from the Sons of Gwalia at Leonora to mines in China. Later he managed the outback Lancefield mine, beyond Laverton, and spent a term on the Lake View itself. By the late 1920s he was living in London as a full-time director of Consolidated Gold Fields, becoming its chairman in 1929. He authorised the exploration which discovered the huge goldfield on the western Rand, one of the momentous events in the economic history of South Africa. Agnew possessed two qualities not normally seen in the one person: he was game to take big calculated risks and yet he knew how to save halfpennies while operating a mine. When, on behalf of his London company, he became chairman of Lake View & Star, he injected the same fine blend of qualities, though injecting them from afar.

Joseph Franklin Thorn became Agnew's new mine manager and then his general manager at Lake View & Star. Quiet and hard-working, Thorn was not much seen at the grand social occasions when Kalgoorlie, during lean times and fat, danced far into the night. For many years it was not quite realised that this American newcomer was a silent dynamo. He was too big physically to be unnoticed, and yet for some reason his talents were not as widely recognised as they should have been. He was the quiet giant of the Golden Mile in the 1930s.

As the depression set in and as the price of gold soared, Joe Thorn — with Agnew's encouragement — created virtually a new mining enterprise. His Lake View & Star led the way in almost everything. It became, along with the big mine at Wiluna, the showplace in the gold boom of the early 1930s, a mine that every student from a school of mines longed to see. Lake View set up the first large-scale flotation plant on the Golden Mile, and soon that process became the vital link in the steps which turned hard rock into almost-pure gold. Flotation was a wet process and so was less hazardous for the men who hitherto had inhaled the fine, sharp white dust at the treatment plants.

Thorn was perturbed by Kalgoorlie's high costs of steam power — the power that worked the rock drills, supplied the ventilation in the underground workings, enabled the winders to hoist up the cages of men and the skips of ore, and worked the machines in the treatment plant. Scavenging money for a new power station, and turning from wood to diesel, he generated his own electricity for his mine. This broke the monopoly of the big London-owned company, the Kalgoorlie Electric Power and Lighting Corporation, with its cavernous wood-fired boilers. Lake View's was eventually the largest diesel powerhouse in Western Australia.

By 1930, nearly every mining expert guessed that if another gold boom was to erupt, the promised land in Australia would be the Golden Mile, simply because it was more impregnated with gold than perhaps any area of similar size in the world. Lake View was one step ahead of the experts. In terms of the mineralised real-estate, every inch on the 'main streets' of the Golden Mile was now securely held by existing owners, and Lake View & Star probably had the biggest single share. Through the shrewd judgement of its staff, Lake View & Star acquired valuable ground from other struggling London companies before the rising price of gold made it seem valuable. The Ivanhoe mine with the most ambitious and capacious shaft — the best vertical underground highway on the field — was bought for shares in 1924. Five years later another grand old mine, the Golden Horseshoe, was bought after the geologist Malcolm Maclaren came to the Golden Mile at Agnew's request and observed how strongly the Horseshoe lodes were silently intruding, at considerable depth, into Lake View ground.

Thorn's last big acquisition, the Associated Gold Mines of W.A. in 1934, gave Lake View control of a large patchwork quilt of prime ground along the Golden Mile. The price of these purchases was high — about two of every three shares in Lake View & Star were now held by shareholders of the three purchased mines. The reward was that Thorn's company, hauling ore up half a dozen main shafts dotted along the Golden Mile, became the largest miner of ore the Golden Mile had seen. In each period of 24 hours Lake View & Star now handled twice as much gold-bearing ore as the biggest mines of the pre-war years.

Joe Thorn, himself a mining engineer, transformed the underground workings of the Lake View. All the old ways of doing things were examined and, if they could not justify themselves, were marked for change. The best underground miners were often conservative and, their life being in their own hands, they followed the ways that had served them well in the past. After showing their opposition they accepted the new ways of drilling certain kinds of ground, new ways of sharpening the drills that were quickly blunted by hard rock, and new mixtures of gelignite for different jobs.

Normally the ore had to be trucked in half darkness a long distance from the working places to the nearest shafts. In the main haulage ways Thorn jettisoned the old iron ore-carrying trucks pushed by hand along a narrow-gauge railway, replacing them with locomotives driven by an electric battery and capable of hauling a string of much heavier wagons at a speed of six miles an hour. Many of Thorn's ideas came from the American mines in which he had worked. Others were devised for the special conditions and the distinctive rocks along the Golden Mile. A few were Thorn's pet ideas. An inventor, he already was known for his patented Thorn pump, a heavy high-lift pump used to dewater old workings.

Miners' health and safety he placed high on his list of priorities. One small matter reflected his concern. His mine gave work to ore-breaking contractors and former tributers who, valuing their independence, wished to be their own boss. They drilled holes in the rock in their own way, using explosives and fuses in their own way, and setting fire to the gelignite whenever they were ready. As a result the mine — according to Ralph Anderson, a member of staff — 'was full of fumes and dust throughout the shift'. The proposed remedy, based on close observation of the characteristics of the underground rock and every phase of mining operations, would save money for the company and reward the underground miners. It merely stipulated that all miners should fire their charges at the same times in each shift, either at crib time or knock-off time, rather than when the need or whim suited them. The practice of firing at the same time would let the dust and fumes settle, thus making the atmosphere healthier for the miners' lungs. Thorn eventually won his point. Normally he employed no industrial relations man; normally he relied on persuasion; but he sometimes called on the American he had recruited, Charles Thielicke, alias 'the slave driver'.

Other managers imitated Thorn. In the early 1930s most of those mines that had sleepily survived the Trembling Twenties were invigorated, sometimes under new owners and usually under new managers. Boulder Perseverance (alias 'the Percy'), managed so brightly in the

From mine to mine, new energy went into winning efficiency. Here in 1934, alongside the ever-present poles of firewood stacked high, was the small narrow-gauge locomotive with the exotic funnel — a replacement for the horse and dray. Less noticeable were the new side-tipping, labour-saving wagons built by the Ruwolt works in Melbourne.

THE DECADE WAS UPSIDE DOWN

At BHP's gold mine, the Hannan's North, boys and men are at work at the picking table in the mill.

1920s by Ernest Williams that he was almost forgiven by the stern royal commissioner, Kingsley Thomas, again made a profit under a new general manager, J. E. Manners. At the 'Percy' the dividend reached 25 per cent in 1936, and then climbed to 40 per cent, remaining there for four years. The London company called North Kalgurli (1912) Limited, chaired by the experienced mining engineer C. A. Whitfield, became a winner as gold prices leaped. It worked shallower lodes than those in the neighbouring mines and so it could mine ore at a surprisingly low cost.

South Kalgurli Consolidated, the result of a merger made in 1913, came into its own under the leadership of the brilliant metallurgist F. G. Brinsden: one of the few of the former Ballarat students still to be found in prominent posts on the Golden Mile. The big steel company, The Broken Hill Proprietary Co., sent its scouts to the Golden Mile and bought the most northerly of the large-scale mines, the old Hannan's North, turning it into a steady producer. Another newcomer, ultimately known as the Gold Mines of Kalgoorlie, gained a slender footing on the eastern edge of the Golden Mile and showed high technical skills. Part of that complicated stable of companies that became Western Mining Corporation, it paid its first dividend before the Second World War but did not become Kalgoorlie's top gold producer until well after the war.

Great Boulder, the biggest producer in more than half of the years between 1900 and 1930, had the chance to outshine Lake View & Star. Instead it allowed much of its mine to remain a bazaar for the tributers. Its output of ore, aided by the busy tributers, increased in the early 1930s but was no match for Lake View.

Great Boulder then fell into the hands of Claude de Bernales. A tall charming man of Basque descent and English birth, he had run a foundry and machinery firm in Kalgoorlie before settling in the Perth suburb of Cottesloe. He stood out in every gathering. His physique, charm and personality made women especially notice him. This seller of secondhand machinery dressed like a London dandy. His offices, whether in Kalgoorlie or Perth, were ornate. In clubland in Perth and Melbourne he was a commanding presence. In his heyday he could sell almost anything, as people were too late in discovering. In the gloom of the 1920s, however, he was a ray of hope for Western Australia, and on visits to Canberra in 1930 did much to persuade politicians of all parties that a subsidy on the price of gold would do wonders for Australia.

His tongue was just as honeyed when he moved to London in the early 1930s and became a mining promoter. With his monocle he seemed to look right through London investors. For years they failed to see through him. In London, as the price of gold rose, he floated new gold companies. A tall pied-piper, he led a procession of investors into the land of his imagination. Of his eight companies, none paid a dividend. That would not have mattered if the money had been spent in legitimate exploring — but much of it wasn't.

In 1936 Claude de Bernales became the managing director as well as chairman of Great Boulder. No man from the goldfields had ever before won the main seat on the London board of one of the big Kalgoorlie mines. With the aid of the celebrated name of Great Boulder he engaged in other mining ventures, some of them dubious and some very dubious indeed. In the end he harmed mining by destroying people's trust in it. In July 1939, on the eve of the Second World War, the London Stock Exchange barred dealings in the shares in the companies he had created — fortunately Great Boulder was not one of his creations. Investigations in England by the Board of Trade did not lead to a prosecution but they pricked what was left of his reputation — except in Western Australia, where he kept a loyal following right to the end. His last years were spent in a mansion with floodlit grounds, the White House at Selsey Hill, near the English Channel. He died there in 1963.

Great Boulder needed a Joe Thorn, not a de Bernales. Nonetheless, even before de Bernales was at the helm, the mine became more efficient. It ceased to mine selectively and eventually escorted the last of the tributers up the shafts. It enlarged and rebuilt its mill and added a flotation plant, de Bernales supplying some of the machinery. It again became an impressive mine working on the large scale. Soon it was paying out an average of 25 per cent a year — wonderful to receive but humble compared to the pre-war dividends that had often reached 150 per cent.

Still, any dividend coming from a company of which de Bernales was chairman was hailed in some quarters as a miracle.

Clusters of smaller companies were floated in Kalgoorlie itself. Beneath the verandas of Hannan Street the practice of buying and selling gold shares had not completely died even in the dead 1920s, and now it was quickened. Local stockbrokers were busy again. Boxes of share certificates, ever so pretty with their garish colours and ornate lettering, arrived from city printers. Local promoters emerged from their long period of hibernation. Readers of the daily *Kalgoorlie Miner* noticed that their fellow citizens were being dressed up like Christmas geese, and displayed as directors of new gold syndicates and companies trying to raise money to explore the tiny local gold leases wedged between the producing mines. Thus in September 1932 appeared in print an unusual foursome of directors promoting an infant mining syndicate: Bernard Leslie, the mayor of Kalgoorlie, Jack Hehir calling himself a 'merchant' and living on the Broad Arrow road, and John Joseph Lynch and William Grundt, who called themselves 'mining investors'. The aim, as in the 1890s, was to sell their gold properties at a profit to a London company.

Visitors who were shown the full stretch of the rejuvenated Golden Mile marvelled at the number of mines, a few small and many big. The revival of mining could be heard as well as seen, especially at midday when each mine blew its whistle or siren to mark the 'crib' or lunchtime break.

Gavin Casey centred one of his best-known short stories, 'Whistles at Noon', on the day — maybe real, maybe imagined — when a Kalgoorlie engine driver thought it was noon instead of 11. Accordingly he blew the mine's whistle, 'its hoarse and angry blare sending a message of cheer to a hundred men'.

Along the Golden Mile to the north, one by one, the other engine drivers sitting in winder houses along the Golden Mile mistakenly sounded their mine whistles: one making a 'piercing high-pitched scream', another letting loose its 'full-throated bass roar', and North Kalgurli's modern siren 'sending out a long-drawn moaning howl to assault the ears of everyone for two miles around'. And so the irregular chorus of whistles marched briskly north until it reached the Northern Deeps (presumably Hannan's North) which finally ended the music with its own 'thin screech'. In the streets of the towns the sound of the 'noon' whistles blowing automatically let the children out of school and sent townsfolk hurrying home for their midday meal. Everywhere people looked at their watches in puzzlement, assuming that they must be wrong. 'Several thousand clocks were taken off shelves, wound, shaken, stared at suspiciously.' Casey's simple story caught the atmosphere of the Golden Mile and the way the mines and the rumbling mills and the commanding whistles dominated daily life.

Along the skyline in the 1930s arose more smokestacks and more poppet heads, their steel or wooden legs standing stiffly above the mine shafts. The main moulders of the skyline were the flat-topped dumps consisting of the dry slimes and sandy residues discarded from the treatment plants of the past. That skyline of dumps was altered with the fresh attempts to extract the last pennyweight of gold from those mountains of old waste — gold that had defied earlier methods of treatment. Families who for years had seen the morning sun rise above the man-made hills of waste saw them change shape and shrink as the sands were carted away for yet another round of processing at the new retreatment and cyaniding works established along the Golden Mile. The wastes from these latest ventures in scavenging were pumped a few miles away to form yet another chain of man-made sandhills.

The increasing scale at which the new ore was mined and treated led to the creation of more steep hills near the Golden Mile. These big 'putty-coloured dumps', as Gavin Casey described them, were the unavoidable results of the mining of gold on a large scale. Even in the richest mine far less than 0.1 per cent of the material consisted of gold, and so the barren 99.9 per cent of waste matter had to be stored somewhere. Normally it was dumped on the surface where on blowy days the wind whipped off the top grains. One quiet change from the 1940s was the use of the waste sands to fill the vast network of underground spaces from which ore had been freshly mined. The dumps ceased to grow at the pace expected. The Golden Mile, however, retained its strange walled appearance.

THE DECADE WAS UPSIDE DOWN

In 1937, cars and utilities are prominent on the mining leases. Here at the Oroya lease the new company, Gold Mines of Kalgoorlie, is building its treatment plant. On the horizon can be seen the steep cliffs of the old tailings dumps which were again being cyanided to extract the grains of gold that had eluded earlier treatment methods.

THE GOLDEN MILE

Another view of the man-made cliffs, near the Oroya South shaft in 1936.

In the unprecedented scale of operations on the Golden Mile, Lake View was the giant. In the year 1937–38 it mined 620 000 long tons of ore — making it the biggest underground mine in Australia. Great Boulder was slower to break records. It had mined and treated more than 200 000 tons of ore in every year between 1909 and 1915, but it was not until 1937 that it returned to those peaks. In the early years of World War II, Great Boulder reached 400 000 tons a year and in 1957 it was to pass 500 000 tons for the first time. The advantage of these high tonnages was that the cost of mining and treating each ton fell away, and very low-grade ore could be mined at a profit. Here was the mining equivalent of the Chicago canneries that were said, when that city was 'hog butcher to the world', to make a profit even from the squeal of the livestock.

While the tonnage mined was a record the output of gold was not. Kalgoorlie's annual output of gold did not again reach the peaks of the early 1900s. The capacity of the field to revive was the surprise. Such was the new production that in the mid 1930s Kalgoorlie's total output of gold caught up to Bendigo's. The exact year when Kalgoorlie became the greatest goldfield in Australia's history cannot be identified with certainty, for Bendigo's statistics are incomplete and consist of some 17 million ounces of recorded gold and estimates of unrecorded gold as high as 5 million ounces. But even if Bendigo's total gold is reckoned at a high 22 million ounces, that tally was exceeded by Kalgoorlie in about 1936.

Kalgoorlie, while remaining the dominating goldfield of Australia, was less overwhelming in its supremacy. Thus for the year 1939 it had Australia's and the state's two top producers of gold, the Lake View and the Great Boulder, but Wiluna and Big Bell held the next two places, followed by North Kalgurli, and then Sons of Gwalia. Thus three of the big six lay outside the Golden Mile and five of the big ten lay outside it. Of the mines on the Golden Mile, North Kalgurli was probably the darling of the investors. Its dividend was 62.5 per cent in 1935, leaping up to 100 per cent in 1937.

In retrospect the 1930s formed a golden age for most of the people on the field. The rest of Australia was worse off than in the 1920s but Kalgoorlie and Boulder were better off. Wages were high, especially in the mines. Hours of work were also improved after the six-weeks' strike during 1935, the only serious strike in the field's history. By 1934 gold was again the main export from Western Australia and the revenue from gold filtered into every corner of the state. Virtually every citizen shared in it, because it made their taxes lower and unemployment lower than otherwise would have been the case.

For one group of people working on the Golden Mile the 1930s were not so secure. 'The foreigner question' did not go away. The question of how many southern Europeans should be allowed to find jobs on the goldfields was periodically debated in parliament in Perth, with the Labor Party calling for restrictions. In the hot summer of January

THE GOLDEN MILE

In 1993 the Western Mining Corporation, under Sir Arvi Parbo and Hugh Morgan, is one of the giants of world mining but in 1934 it was an infant even in Western Australia. A bold infant, it conducted an aerial survey of parts of Western Australia in the hope of finding orebodies. This aerial view of a corner of the Golden Mile in 1934 shows the boomtime miners altering the bare landscape with new mills, tailings dumps, railways, roads and more dumps. The craters and scars give the impression of an industrial centre that has just been bombed.

1934 the issue spilled onto Hannan Street. On Monday 29 January, a popular footballer died after fighting an Italian-born barman on the pavement outside the Home From Home Hotel. On the following evening several hundred men, mostly young, wrecked the two-storeyed stone hotel and then set alight the nearby wine saloon and wrecked the All Nations Boarding House — places favoured by Southern Europeans. The mob, now larger, moved along Hannan Street, wrecking and looting chosen shops, before taking over trams and travelling to Boulder where the men selected more targets. Hundreds of foreigners left the goldfields or hid in the scrub. By the time the massive reinforcements of police had arrived by train from the coast, the damage was done.

The mines were closed for nearly a week. Mass meetings of miners called for a ban on unnaturalised Australians being employed on the Golden Mile; and eventually it was agreed that only miners speaking the English language should be employed. The Premier, Collier, himself an old Boulder miner, rightly regretted that such a 'scandalous thing should have happened in a civilised community'. A stain rightly remained on the name of the goldfields.

There has been a recent tendency to moralise too eagerly about the behaviour of Kalgoorlie in 1934 and the character traits of mining towns. Even larger mobs have been known to assemble in the big cities. There, however, the police could easily be mustered in larger numbers to restore order before the danger or damage to life and property became acute. In contrast the few police on the goldfields, during the many hours of crisis, were almost powerless. There was no hope of reinforcing them quickly. Likewise the grievance about employment was inevitably much higher in a one-industry town. As the main source of work was mining, and as many foreigners were employed in the mines, it was hardly surprising if local residents felt especially aggrieved at losing their only opportunity of work.

This observation does not condone the deplorable behaviour of the mobs on the rampage on those hot evenings of 1934. But it tries to point to those factors which, operating in remote mining fields, were somewhat overlooked by policy-makers of the day, were ignored by moralisers of later days, and therefore could be repeated if similar conditions should again arise.

Slavs and Italians had played their part in the life of the goldfields. For decades they had worked hard in the mines and woodcutting camps, adding to the nation's wealth. Their plight in 1934 was unenviable. A solution, a compromise, to the polarised arguments on both sides was not easily found, but slowly the relations improved. Most of the long-standing foreigners continued to work on the Golden Mile but their ranks were no longer enlarged by newcomers from Italy, Greece and Yugoslavia.

The Italians, but not Greeks and Yugoslavs, again became vulnerable early in World War II. In May 1940, with Mussolini expected to join Hitler's side, lists were made of

THE GOLDEN MILE

Policemen assemble in plain clothes with just a little luggage, in readiness to board the train to carry them to riot-troubled Kalgoorlie in January 1934. (Photo from West Australian Newspapers Limited)

all the foreign citizens working on the Golden Mile. The Paringa mine had none. North Kalgurli employed 12 foreigners who were naturalised but four 'Slavs', three 'Jugo-Slavs' and one Italian miner remained foreign citizens. The Gold Mines of Kalgoorlie was not sure how many foreigners worked for it. Maybe they numbered 22, including four foreigners — presumably Italians — who worked in the tribute parties. South Kalgurli employed two Australians of Italian ancestry and five Italians who were naturalised and three other Europeans — a Norwegian, a Swede and a Yugoslav — who were not naturalised. While the lists from the two biggest mines are lacking, clearly the Italian miners must have been fewer than in 1934.

At two in the morning of 11 June 1940 the news reached the goldfields that Italy had entered the war on Hitler's side. Several hundred unnaturalised Italians, hurriedly detained, were sent by train from the Sons of Gwalia mine at Leonora, the stronghold of the Italians ever since Herbert Hoover had managed the mine back in 1900. According to the *Kalgoorlie Miner*, the Italians 'proved most docile', some readily giving themselves up. Placed under armed guard in Kalgoorlie, they were presumably joined by the small number of Italian citizens working on the Golden Mile. After internment on the island of Rottnest, some men were soon released.

The Australian Workers' Union persisted with its old peacetime argument that those who spoke no English should not be allowed to work in the mines. In June 1940 the Chamber of Mines representing the mine-owners agreed that even the use of a foreign language should be banned in the mines. It was easier to ban foreign tongues than to permit resentment to simmer in the mines.

However, a big old underground mine is probably the most difficult workplace of all for the censoring of speech. Most of its men work in groups of two or three or four, scattered in remote places which only the foremen or shift boss visit — and then perhaps only once — in the course of a shift. Even when a boss approaches them, the loud staccato noise of the rockdrill and the sound of the compressed air combine to blur the spoken word. No doubt those Italians and Yugoslavs who happened to work together in small teams in remote workings continued to speak their own language while at work.

The Golden Mile was reaching the peak of a wonderful recovery when the war began but by 1941 the momentum had gone. After Japan entered the fighting in December 1941 the nation's gold was not deemed vital for the war effort, and hundreds of miners were allowed to enlist or were diverted to essential industries. The gold mines had to give up the vital task of developing new underground areas for ultimate mining, and Lake View & Star did a mere 6000 feet of development work in 1943 compared to 40 000 feet in 1940. Record tonnages were replaced by low tonnages. The monthly output of gold slumped. Thus the Second World War was to be even more damaging than the First to the Golden Mile.

SOLE SURVIVOR

CHAPTER NINE

AFTER THE WAR, Kalgoorlie's route to recovery was longer than expected. Cement and steel, explosives and other raw materials remained scarce. Miners also remained scarce, though they were allowed to return to the mines from wartime jobs. The reduction in 1947 of the working week to 37½ hours for miners and 40 hours for surface workers meant that more labour was needed but in compensation these improved working conditions helped attract new miners.

In February 1948 a deluge of 12 inches of rain in an hour — more than the rainfall of the average year — poured silt and water down shafts, flooding the lower levels of several big mines. But mishaps were matched by discoveries. In the old Iron Duke leases of the Golden Mile such rich veins of gold-telluride ore were found that steel strong-room doors had to be built to guard the entrances to that lode called the 324 Series and so reduce the theft of gold. Only the shift bosses were entrusted with mining that precious ore. It was so rich that even the dust was swept, placed in sealed bags and sent to the mill.

An ailment that was almost hereditary in the economics of gold mining was becoming visible again: the steady increase in wages and other mining costs was not paralleled by a rise in the price of gold. This was the rheumatism of the gold industry. It seemed to defy all remedies.

Miraculously a temporary cure for the condition was announced in 1949. The British government devalued the English pound — a repeat of one of the dramatic financial events of the 1930s depression. The price of gold in Australian currency jumped by nearly 50 per cent, and hovered just above 15 pounds for the best part of the next 20 years. At first that was wonderful news. Hannan Street rejoiced. But the cheers soon faded. In the early 1950s the Korean War brought the worst inflation experienced in Australia since the gold rushes of the 1850s, wiping out in a couple of years the gains from dearer gold.

Shrewd observers thought that Kalgoorlie would quickly rejoin the frustrating treadmill of the 1920s. Hope, however, came from Canberra. During the federal election of 1954 the Prime Minister Sir Robert Menzies promised that a new gold bounty, not unlike that of 1930, would help to keep the goldfields alive. After winning the election he made the bounty even more generous. In the federal parliament on 4 November 1954 he personally introduced the Gold Mining Industry Assistance Bill with an affirmation that gold was still important to the balance of payments because nearly all was exported. Gold was also vital for the remote parts of the continent, especially Western Australia. It would not be in the nation's interest, he said, if those areas were depopulated.

The bill was mainly a law to help Kalgoorlie, once again the mainstay of gold mining in the nation. Indeed H.V. Johnson, the former AWU organiser who was the member for Kalgoorlie, opened the debate on behalf of the Labor Party, giving his support to the bill. It promised a subsidy, of up to two pounds for each fine ounce, for those mines that suffered from high costs of production but were willing to persist. If the cost of producing the average ounce of gold was more than £13 10s, then the mine was entitled to a federal subsidy. The subsidy equalled three quarters of the excess cost of production. Thus a mine with working costs of 15 pounds an ounce received a subsidy of £1 2s 6d. The maximum subsidy was to be two pounds an ounce. The scheme, it was promised, would last two years. In fact it lasted for about 21 years, being renewed or amended on many occasions. In the event it did not prove to be a prodigal scheme. Sir Lindesay Clark, who was probably the main lobbyist for this scheme, later computed that in its whole life it equalled a subsidy of about 6 per cent on the cost of every ounce of gold mined in Australia during that period. In other words, this gold subsidy was a drop in the ocean compared to the subsidies received by the typical Australian factory.

Curiously, the dramatic boost to the efficiency of the Golden Mile came not from a debate in parliament but from the ingenuity of men working underground and on the surface. A quiet but sweeping change, begun in the 1930s, was accelerated in the 1950s. The method of mining was never to be the same again. Whereas sheer strength as much as skill was the basis of old-time mining, the new regime of the late 1950s depended more on better machines and techniques than on muscles. Fifty new ideas and devices were at work, each of them saving labour and some of them increasing safety.

Underground, the miner now wore a cap-lamp lit by electricity instead of the smelly carbide lamps. He drilled holes into the rock with 'airleg' machines that were lighter, faster, and easier to move to another position: many of the old mechanical drills were so heavy that they could only be moved with the strength of two men. At the point of the machine the drill itself was tipped with tough tungsten-carbide, giving it more penetrating power and a longer life. Indeed nothing did more to prolong the life of Kalgoorlie than this Swedish innovation that arrived around 1950.

In blasting the newly-drilled holes, explosives of greater power and higher safety were used, culminating in the 1960s in the ammonium nitrate mixed with fuel oil. In handling the broken ore, the miner ceased to depend mainly on his shovel and muscles and instead used a mechanical loader or 'bogger' that saved energy and gained speed: the first boggers had been tried in the 1930s. In sending the ore from the workplace to the lower transport level, the iron ore-pass lasted much longer than the old wooden pass. Along the main underground levels or passageways that led from the ore-breaking areas to the shaft, small battery-operated locomotives took the place of the man pushing the truck full of ore.

The mined ore was hoisted up the shaft by powerful winders all now driven by electricity rather than steam. In the engine houses the huge flywheels belonged to the past. Traditionally the headframes or poppet heads that

THE GOLDEN MILE

Underground the battery-operated locomotives replaced the men pushing the trucks along iron rails. A warning bell was close to the driver's right hand in this mine.

stood above each shaft, with small wheels revolving at the top, had been made from forest logs, but now the tallest headframes were made of steel. Several had been dismantled and carried from closed gold mines on the Murchison. The noblest poppet head ever seen in Kalgoorlie arrived from Big Bell in the Murchison, like the parts of a Meccano set, and was erected above the Ivanhoe shaft. It is now a landmark and lookout tower in its new home, the Museum of the Goldfields.

The 1950s especially gained from these novel machines and new ways of mining and hauling ore. The output of ore from each underground worker went up year by year. Underground, about 250 men were now doing the work that 1000 men had done with enormous effort at the start of the century.

These changes were remarkable because they had to be accommodated inside a vast rabbit warren, a warren created on an astonishing and majestic scale by decades of busy work. Thus in 1960 more than 300 different orebodies were being worked on the field, at varying depths. Great Boulder, working 70 different lodes, was gaining some very rich ore at a mere 400 feet below the park opposite the Fimiston Hotel but it was also sinking an internal shaft at a depth of 3850 feet or more than seven tenths of a mile below the surface. The system of underground transport by which the ore was sent horizontally and then raised vertically was inevitably more the result of unforeseen events than of planning: nobody knew about more than a fraction of the orebodies and nothing about their depth when the system of openings for transport was first laid. It was extraordinary that by 1960 about 150 different locomotives of small dimensions were fitted into this warren. Ideally new shafts were needed but they were too expensive for a field that could not be sure of its future. Only the brave New Kalgurli under Alec McLeod sank a major shaft in the 1950s, enabling large teams of men and quantities of ore to be quickly carried up and down.

Much of the new efficiency depended on altering machines to suit Kalgoorlie's special needs and the kind of spaces that already existed in the underground mines. Some of this efficiency came from using scrap machinery and scrap metal to make or remake equipment. A goldfield so old, with a procession of technical changes, had become one of Australia's most fascinating junkyards of old machinery, much of which was kept in the hope that some parts might still be useful. Colin Yates, the number-two man to Edgar Elvey at Great Boulder, was asked in 1961 if his company ran a salvage department. 'Yes,' he quipped, 'we have a good Scotchman on our staff: he also kills the white ants.' To which a bystander replied, 'Great Boulder has dozens of Scotchmen'. Every mine by then was skilled at improvising.

Above all, the Golden Mile retained a host of first-class workmen. Many men took to new machines with ease. In attitude the miners and mill men were less conservative than a generation ago when many had dug in their heels at the thought of change. Peter Wreford, who began as an underground labourer at Lake View in 1939 and later became one of the country's best-known mine managers, recently paid his old mates at Kalgoorlie the highest of compliments. He said they could work well as teams. The highest-paid miners had a 'real instinct for their craft'. Relations between foremen and miners were reasonable, despite the abusive language that might pass between them. Nearly all men took pride in working hard — 'bludger' was a word of stark disapproval. Absenteeism was not common, and frowned on by workteams.

A man might go to the surface after receiving a gash on his arm or leg; the first aid man would look at the injury and send him to his doctor to be stitched up. If the gash was simple, clean and not extensive, the miner would be back at work on the next day.

A miner who stayed away from work for another day, without valid reason, was looked on by his own mates as a 'booby'. Wreford recognised that the Golden Mile owed much to those who worked there, often doing exhausting work under trying conditions. Never again, Wreford wrote, did he come across a workforce with as many people who were 'cheerful, easy going, hard working and all round sensible'.

On the extensive surface works a complex chain of processes extracted the quarter ounce of gold from each ton of barren rock. Here the technical advance was not so rapid, but advance there was. More efficient flotation machines were built. New reagents helped to separate the grains of minerals from the unwanted grains. Small savings were made here, new ideas adopted there, the big advances having been made in the 1930s.

The new leader in metallurgy was Reg Buckett, a young South Australian with a Moonta background on his father's side. He completed his metallurgy course at the South Australian School of Mines in December 1930 — 'a grim time to be looking for a first job,' he recalled. Seeking work at Kalgoorlie, he spent six months at Great Boulder in the humble post of assistant to the mill fitter. The next months were disappointing, before he found a job working as a mullocker at the Croesus shaft. By the standards of those hard years, when one in four Australians had no work, his progress was middling. Buckett's chance came in 1932 when he was asked by Joe Thorn to operate the laboratory at its Chaffers treatment plant, a home of the flotation process on the Golden Mile. Three years later, he was appointed mill superintendent. To be placed in charge of the main treatment plant of the biggest gold mine, with perhaps the biggest tonnage of any metal mine in Australia, was a thrill for him. As overseas experts had not improved the mill enough, the burden was placed on this stripling in his mid twenties. Some of the old hands were not pleased to see such a young boss. But bit by bit he showed them how to improve the mill, making it probably the most efficient and biggest in Australia. Eventually,

SOLE SURVIVOR

In the Great Boulder in the early 1960s the drill with the tungsten-carbide tip was transforming the art of mining.

other mills along the Golden Mile were not far behind.

Cheap power was another success, with coal and diesel fuel slowly replacing firewood. At the end of the Second World War, however, firewood-cutting was still a big industry in the scattered scrub far beyond Kalgoorlie. Larry Hunter recalled in his booklet *Woodline* how in 1946, being posted from Perth to teach the children of woodcutters at the Lakewood Main Camp, he reached Kalgoorlie by train only to learn that he still had to ride seven miles in a taxi and then, in the darkness of early morning, travel 59 miles in a narrow-gauge tram. With such long haulages it was not surprising that the price of firewood, multiplying by four and a half times between 1940 and 1958, was no match for other fuels.

At Boulder the Kalgoorlie Electric Power & Lighting Corporation clung, for too long, to firewood as fuel. When, after half a century of burning firewood, it changed to coal in 1953 at massive expense, the majority of its big customers had already said goodbye. Most mines now generated their own electricity, using diesel fuel. Ten years later the big London-owned company let its fires go out for the last time.

The field was breaking new records, not in profits but in sheer output of ore. In 1960, Lake View & Star mined and treated 760 000 tons of ore, or three times as much as the biggest company had treated in the rich years before the First World War. For the time being, sheer efficiency and large-scale production of low-grade ore were substitutes for the fact that the richest ore seemingly had all been mined. Meanwhile the costs were creeping up, while the price of gold remained the same. Edgar Elvey, who ranked with Richard Hamilton as one of the outstanding leaders of the field, confided in 1961 that he would be the last general manager of Great Boulder. He could see that the old mine had less than one decade of life ahead of it. Already it had produced a grand total of nearly 6 300 000 ounces of gold and had paid a return of 4474 per cent on the original capital. Its warren of workings still held plenty of ore: to make a profit was the problem.

A latecomer, the Western Mining group, had now ousted Great Boulder as the second most productive company on the field. The group was born in 1930, on the eve of the gold boom of the depression years. Its real founder was the Melbourne-born businessman W. S. Robinson, one of the kings of international mining in the years between the two world wars, and its biggest financier in the first year was John Agnew's Consolidated Gold Fields, although he later withdrew.

In opening gold mines around the continent, the Western Mining group had an early mixture of success and frustration. Its early successes included Mount Coolon near Bowen in north Queensland and Triton in Western Australia; its long-delayed successes included Central Norseman which became, after many obstacles, one of the most rewarding of gold mines. Its remembered failure was a grand attempt to revive the Bendigo goldfield, less than 100 miles from Melbourne.

Meanwhile it gained a firm foothold at various points on the Golden Mile. There it operated mainly under the name of Gold Mines of Kalgoorlie (or GMK), a company floated in London in 1934 though its head office was transferred to Melbourne after the war.[1] It arrived in Kalgoorlie too late to obtain really promising leases; but with the aid of outstanding geologists it made the most of its ground, mining ore from the old Iron Duke and South Oroya shafts. In 1937 it paid a maiden dividend — four and a half pence on each ten-shilling share. Its mine was managed by Western Mining, using the Kalgoorlie offices that had once been occupied by the famous consulting firm Bewick, Moreing. To that office came a procession of talented engineers, metallurgists, and especially geologists trained in America. Its special contribution to the Golden Mile was new blood and new ideas out of all proportion to its importance as an actual gold producer.

Western Mining Corporation employed outstanding superintendents at Kalgoorlie, including old Lou Westcott, young Keith Cameron, and then 'Father' Frank Espie. After the Japanese army invaded British Burma, Espie had bravely led a party from his mine in remote northern Burma all the way to safety in India. Espie brought with him his Burmese servant, Sammy, who in 1943 became one of the conspicuous figures in Kalgoorlie. With cheerful dignity Sammy waited at table in Espie's dining room, looking after the dignitaries who visited, and then on Sunday night stood with the Salvation Army and its band at the intersections in Hannan Street.

In its early years the Western Mining group was ranked far below the top producers on the Golden Mile, though it was eager to innovate in technical matters. To the surprise of the locals it eventually became the evangelist for the Golden Mile, believing in its future when other companies were becoming lukewarm. Its technical and later its financial leader was G. Lindesay Clark, the son of a well-known Tasmanian mining engineer. Clark was tall and quietly spoken, winner of the Military Cross on the Western Front, a romantic who was fascinated by the search for minerals. He retained faith in gold when most economists thought it had lost forever its vital role in finance. Between 1952 and 1954 Clark's Western Mining group — operating though the company called GMK — tried to buy up as much as possible of the Golden Mile,

---

[1] The sequence of company births in this group was complicated. The first company in what became the Western Mining group was Gold Mines of Australia, formed in 1930. Western Mining Corporation itself was formed in 1933 to explore gold in the western part of Australia. In July 1934, Gold Exploration and Finance Co. of Australia was floated as an umbrella company, holding most of the shares in the earlier company. In October of the same year Gold Mines of Kalgoorlie was floated, taking over the Kalgoorlie leases and options owned by the group and formerly held in the name of the Champagne syndicate. Much later Western Mining replaced Gold Exploration and Finance as the controlling company.

buying out the Paringa, the reborn Perseverance, South Kalgurli Consolidated and a large share in Kalgoorlie Enterprise. Clark was the most diligent lobbyist for the gold subsidy, conferred in 1954 by Sir Robert Menzies. By 1960 the Western Mining group under Clark was the largest gold producer in Australia, and on the Golden Mile its local company, GMK, was second only to Lake View & Star.

The Western Mining group, even before it became the nation's leading miner of gold, possessed an unusual quality for those days. It believed in geologists and it believed in searching. In the mid 1930s the group employed more geologists with experience than probably all other mining companies in Australia added together. It believed that imagination and geological theory had a special role in mining, and Clark happily subscribed to that idea. In 1950 the group formed a company, Kalgoorlie Southern Gold Mines No Liability, with the aim of making the boldest search in the history of the Golden Mile, and began to drill deep and expensive holes in the hope of finding a new gold lode that might lie buried to the south of Boulder township. 'The project,' Clark reminded shareholders in 1959, 'was begun as a highly speculative attempt to find a great prize. The prize is the repetition of the Kalgoorlie gold field.' When people pointed out that if found it might turn out to be just a huge low-grade orebody too deep to be payable in the sobering economics of gold mining, he replied that if it was a real replica of the Golden Mile its first 10 000 000 ounces of gold would come from ore averaging 17 pennyweights to the ton. It was a compelling argument in some minds. The brave and costly attempt had its moments of excitement during nearly one quarter of a century of intermittent drilling. Some holes were nearly 7000 feet long or about one and a half miles into the rocks. But the second Golden Mile was not found.

At the northern end of the field, right at the end of Hannan Street, another bold project was contemplated from time to time. This was the Kalgoorlie end of the field, the neglected end except during the first exciting months of the field. It was separated from the Golden Mile by huge blocks of barren and displaced ground, including the Golden Pike Fault. The only successful northern mine was Hannan's North which, under the control of the great steel-maker BHP, was mined on a moderate scale for 17 years until its closure in 1952. In the course of its life it had averaged about nine pennyweights to the ton, a pleasing result. But its total output of gold, about 275 000 ounces, was no more than the biggest mines on the Golden Mile produced in just one boom year. The other promising but frustrating northerly mine was Mt Charlotte.

The Western Mining group had long thought well of the old Mt Charlotte and Hannan's Reward leases that occupied the ground where, in 1893, Paddy Hannan had first found gold. In 1937 one of Western Mining's effervescent team of geologists, H. J. C. Conolly, privately said

SOLE SURVIVOR

Kalgoorlie Southern, drilling in the late 1950s on the shores of a salt lake in search of a second Golden Mile.

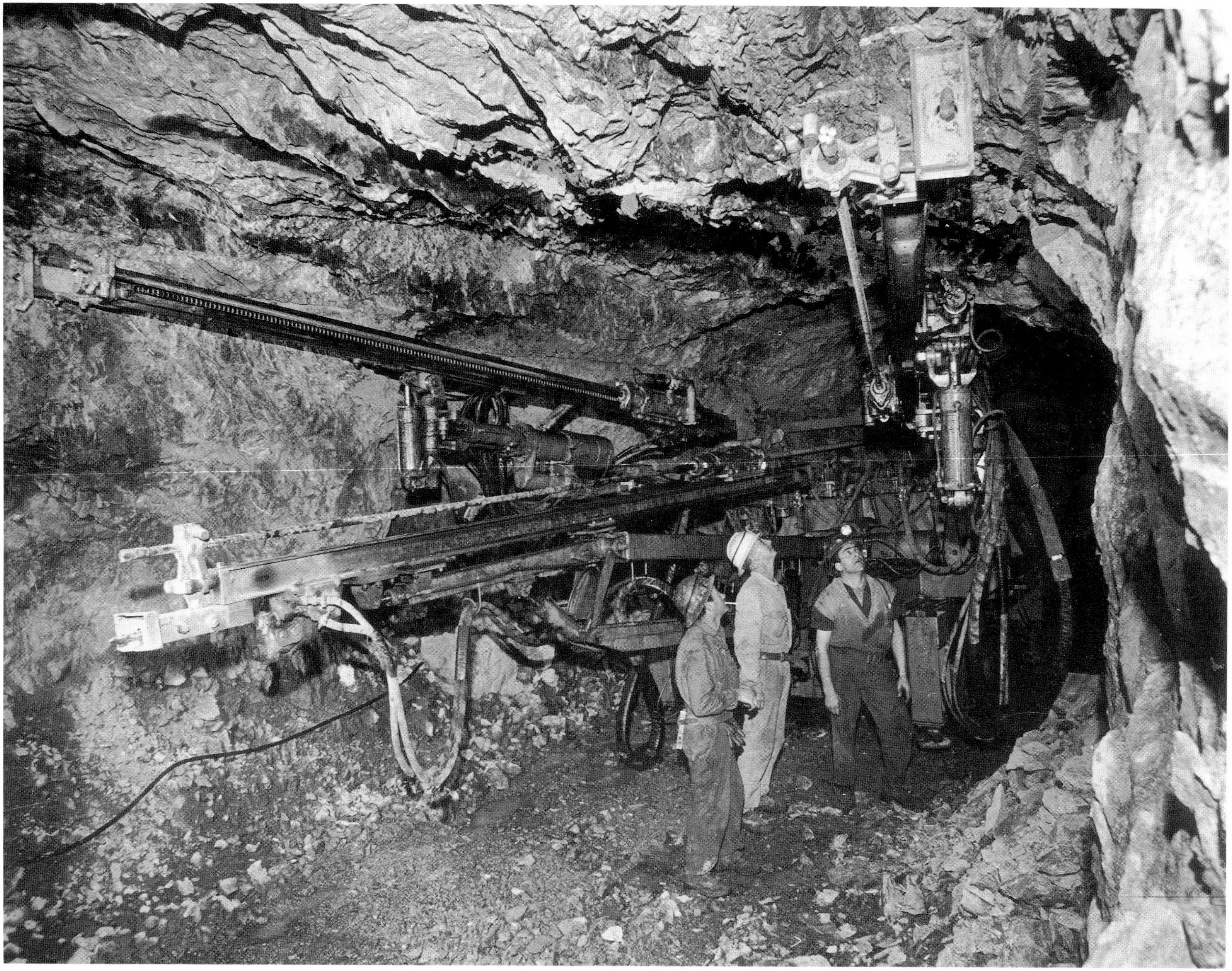

SOLE SURVIVOR

From the 1960s the Mount Charlotte underground mine steadily introduced a new era of mechanisation. The working space in early underground gold mines, even the biggest mines, had been cramped: but at Mt Charlotte the spaces were designed to admit the big rubber-tyre machines that each did the work of dozens of men: the new jumbo rock drill (opposite) and the new version of the long-handled shovel (right).

that Mt Charlotte had the potential to become Australia's leading gold producer. A quarter of a century later, the Gold Mines of Kalgoorlie bought the leases and began to prepare the orebody for large scale mining. It was very low-grade, with a likely three pennyweights to the ton, but it was massive compared to the numerous narrow and deep lodes along the Golden Mile. Fortunately, Mt Charlotte lent itself to mining on the large scale. Exploration revealed that it was the largest orebody on the field, being about 700 feet long and more than 150 feet wide in most places. Above all it seemed likely to go down and down.

Mt Charlotte in the 1960s became the most mechanised mine on the field — the first underground mine to use a squadron of diesel vehicles running on heavy rubber tyres. It brought a new continuity to Kalgoorlie at a time when most of the news was disappointing. Moreover it soon turned the Western Mining group, that late arrival of 1930, into the biggest producer on the Golden Mile.

Always exploring, the Western Mining group made another major discovery which, ironically, seemed likely to terminate the long life of the Kalgoorlie goldfield. At Kambalda, less than 40 miles to the south, was a small goldfield worked at the start of the century. Abandoned, it still attracted prospectors, two of whom found a little nickel on the surface. Skilled geological work by Roy Woodall of Western Mining led to a drilling campaign. In January 1966 the drills passed through rich nickel. Here beneath the dry hills and salt lakes was one of the world's important nickel fields. The mine was developed; power was at first transmitted from the GMK powerhouse on the Golden Mile, and the town of Kambalda was built in a park-like setting. A big nickel smelter was completed in 1972, its tall chimney smoking within sight of Kalgoorlie; the refinery was already opened at Kwinana on the coast; and within eight years of the discovery of the first lode Western Australia was producing about one twelfth of the world's nickel.

The nickel was wonderful news for Kalgoorlie but not for its gold mines. Kalgoorlie became the centre of a feverish mining boom and prices of property soared. Gold, however, had lost its glamour. The gold mines, struggling for survival as gold prices remained firm but costs of mining went higher, were buffeted. They lost many of their best miners as well as a host of clerks, surveyors, mill men and other occupations. Those who remained at the gold mines called for higher wages. Even the surviving companies became more interested in nickel because it offered the hope of high profits at a time when gold mining seemed doomed.

Great Boulder and North Kalgurli, two of the four surviving mines on the Golden Mile, found nickel deposits in the bush beyond Kalgoorlie and began to think of treating nickel ore instead of gold ore in their treatment plants. In December 1969 the most productive gold mine in the history of the field, the Great Boulder, ceased to treat gold ore in its plant. Two months later the second of the giants, Lake View & Star, announced that all development of new gold lodes would cease and the company, with a declining workforce, would clean up the remaining ore during the next three or four years and finally close the mine. It seemed certain that the end of the Golden Mile was in sight, leaving only the Mount Charlotte mine to keep the field alive — if it could survive the rising costs of production.

As if to signal the end, the Chamber of Mines of Western Australia moved its head office from Kalgoorlie to Perth in 1970. A year later its president, L. C. Brodie-Hall, executive director of Western Mining and a fighter for the goldfields, summed up the crisis: 'The gold mining industry is rapidly approaching its end.' The challenge was to keep a few gold mines alive at Kalgoorlie a few years longer and so keep that big town and all its facilities alive until such time as further base-metal discoveries would take up the running. Even that short-term task seemed impossible. The unionists, conscious that they were lowly paid compared to nickel employees, called for another $16 a week. The surviving gold companies could see the end in sight and decided the time had come to close their mines in an orderly manner. All they sought was five or six months in which they would mine the ore that was already blocked out for mining. In those months the workforce would fall away until none remained on the payroll. Even that slight reprieve, at the new rates of pay demanded, would cost another $640 000.

One Friday afternoon in Perth, Brodie-Hall called on the Labor Premier, John Tonkin, and said that four big gold mines — including the three survivors on Kalgoorlie — would probably have to close down. Tonkin promptly placed the matter before cabinet. On Monday evening he rang Brodie-Hall to say 'You've got your money'. The four mines scrounged half the cost of the pay rise and the government's grant of about $320 000 met the other half.

The three surviving gold mines kept themselves alive, sometimes by narrow margins. They ceased to explore for ore and ceased to develop new blocks of ore ready for mining. They simply cleaned out the ore in sight. They were rather like farmers who harvested for the last time and had no intention of ploughing in order to sow next year's crops. They had no alternative but to give in.

Nickel continued to boom, using its plentiful money to draw employees away from the Golden Mile. In 1971 for the first time, nickel employed more people than gold. Two years later the three survivors on the Golden Mile were virtually reduced to two when North Kalgurli converted its treatment plant from gold ore to nickel ore. The small tonnage of gold ore it mined was sent to a nearby treatment plant.

Of the major mines that had once produced gold, two only remained. One was Lake View & Star, now owned by Poseidon Limited, a nickel producer. It was a sign of the dulling lustre of gold that Poseidon had bought out this once-famous mine not for its gold but for its treatment plant. The other survivor was Gold Mines of Kalgoorlie, the main gold producer. In May 1973 the two survivors sensibly amalgamated, partly to cut costs but also to work together to revive the field should gold prices increase. The new company, the sole survivor, was called Kalgoorlie Lake View. Each of the partners received 47 per cent of the shares in the new company, while the remaining 6 per cent were purchased by Western Mining Corporation, which would manage the mines.

Here was the most unifying step in the field's history. The separate mining, clerical, metallurgical, geological, purchasing and other divisions of the two companies were brought together. Money was saved. With the price of gold favourable in 1973 and 1974, new ore was developed, old shafts and rattling plants were updated, and plans were made to integrate operations. Kalgoorlie again had a future, everyone said. In the remarkable international financial events of the early 1970s — the end of Bretton Woods, a rate of inflation not surpassed in this century, an oil crisis — the commodity of gold was suddenly liberated. Its price was allowed to move freely on the open market. And at first gold leaped to record levels, but much of this gain was lost to inflation.

For a short time the price paid for Kalgoorlie's gold was more than four times as high as in the 1960s. On 3 April 1974, gold reached a new high point of $US179 or $A120, before declining again. While the price of gold fell, all costs rose during this most inflationary decade in the last 200 years. All the costs saved by the merger involved in the new Kalgoorlie Lake View meant nothing. Dismissal notices were sent to hundreds of men. The Indian summer for the gold mines seemed to be over.

About 23 000 people were still living in Kalgoorlie but, with the closing of the gold mines, 7000 were expected to leave in the next year or so. There seemed to be no way of preventing such an exodus. In recent years, or almost every year of the last 15, there had been some kind of appeal from mining or civic or union leaders to the governments in Canberra and Perth. Governments were now deaf, and could see little point in propping up the ageing Kalgoorlie.

Every mining field eventually dies. Most omens said that Kalgoorlie's time had come.

CHAPTER TEN

THE GOLDEN HOLE

THE YEAR 1975 was memorable for Australia. The economy was shaky, inflation was swift, and politics were in turmoil even after the constitutional crisis was ended with the dismissal of the Whitlam government by the Governor General and then by the Australian voters. For Kalgoorlie, 1975 was even more bewildering. The goldfield was close to extinction, and it seemed that nothing could save it. Efforts to prop up Kalgoorlie had almost exhausted the funds of the surviving companies and the patience of the governments in Perth and Canberra. The gold subsidy introduced in 1954 was no longer of use. The other valuable federal concession, which exempted gold mines from paying income tax on profits, had long ago become meaningless, because no Kalgoorlie mine paid dividends. At the end of 1975, North Kalgurli finally closed its gold mine. Great Boulder, now mining a small open cut on the Golden Mile, finally ceased work in June 1976. Even the joint venture that was the last hope, Kalgoorlie Lake View, was on the brink of closing down. It could not continue to lose money every week. For years no dividend had been paid from the dwindling gold output of the field.

The leader of the field, Sir Laurence Brodie-Hall, wondered whether he should throw the dice just one more time in the hope that an ounce of luck would prolong the life of the gold mines. He had migrated from England to Western Australia as a young man in the late 1920s. Labouring in humble posts on the land and in mines, he went to the war, then joined the Western Mining group and worked his way up, studying at the Kalgoorlie School of Mines in his spare time. No technical institution in Australia has been more influential in its home town than this school of mines, and Brodie-Hall remained one of its most grateful students.

In the crisis of 1975 he was the chairman of the joint venture, Kalgoorlie Lake View, and he looked around for a source of funds that might enable the field to cling on — until the price of gold was increased or a new discovery was made. Brodie-Hall turned to the biggest American gold producer, the Homestake Mining Company, which had been mining gold in the high hills of South Dakota since 1877. Homestake and the Western Mining group had many links. Several of Homestake's leaders had worked, early in their career, as geologists for the Western Mining group in the 1930s, and the two companies had also combined in an iron-ore project in Western Australia in the early 1960s.

Talks initiated by Brodie-Hall ended with Homestake agreeing to inject up to $A8 million into Kalgoorlie. In return it would receive 48 per cent and Kalgoorlie Lake View would receive 52 per cent of a new partnership to operate the main part of the field. Kalgoorlie Lake View handed over its mines and mills, and its new American partner supplied working capital. The new organisation was given the unglamorous name of Kalgoorlie Mining Associates. Born on 10 March 1976, it lived to see golden days but they were far ahead.

Homestake, as agreed, began to pour in money to meet the losses. It was like pouring water into a hole in a desert. Four months after Homestake's entry the price of gold fell to $83 an ounce, partly because the International Monetary Fund was selling some of its gold stocks. Kalgoorlie seemed to have no future. Most employees were given notice but many left before notice arrived. The decision was reluctantly made to stop the pumps and allow the bottom of the shafts on the Golden Mile to fill with water. Mount Charlotte, now the only producing mine worth talking about, made preparations to close near the end of the year. Once closed, the field seemed unlikely to reopen at any time in the next ten years, if at all.

But the outlook, just a fortnight before the Mt Charlotte mine was to close, suddenly improved. On 29 November 1976 the Australian dollar was devalued, and that instantly increased the price of gold by about $22 an ounce. The worry was that the fast inflation of wages and all other costs would gobble up that increase in gold revenue in little more than one year. Homestake, less optimistic than the Western Mining group, had to be persuaded that the Kalgoorlie field was worth clinging to. On 9 December, one day before the operations were to cease, Homestake agreed to keep Mt Charlotte in production. Even then the employees of that sole survivor continued to decrease, falling below 100 before they increased again.

Slowly Mt Charlotte was revived. It remained the only worthwhile mine in the next six years. To the south the Golden Mile itself, the traditional heart of the field, was virtually deserted. Here was a big town built on gold, with more than 20 000 people still fundamentally dependent on mines, scavenging a living from a host of activities of which gold was now one of the weakest. The smoke of the nickel smelter to the south had become the main symbol of the field.

The handful of men still holding senior posts in Kalgoorlie Mining Associates usually turned, when opening the morning newspaper, to read the latest price of gold, because it more than ever determined the future of Kalgoorlie. To their delight they saw gold creeping up, then shooting up during the Iranian crisis of late 1979. In January 1980 the price was a record, leading to mammoth profits from low-grade Mt Charlotte, the last surviving mine. In 1979 the depression-born company Gold Mines of Kalgoorlie, one of the three partners in the American-Australian alliance that now controlled the field, was able to pay its first dividend for eleven years.

All kinds of projects were brought out of the Kalgoorlie cupboards, for the high price of gold made them feasible. A new crushing plant was built for Mt Charlotte. Even the unwanted discovery that the main orebody at Mt Charlotte had, at depth, been bodily moved sideways by an upheaval occurring in ancient times did not cause the consternation it would have in a time of low profits. Diamond drills began to search for the displaced orebody and found it, some 150 metres to the west and 150 metres below the faulted ground. In 1981 the huge sum of $41 million was allocated to sinking a new circular shaft, the Cassidy, to give access to the displaced Mt Charlotte orebody at depth. It was completed four years later, and its poppet head standing on the low hill was brilliantly lit at night — one of the finest industrial sights in Australia and the beacon of the revived goldfield.

On the old Golden Mile the activity was even more vigorous. At the old Oroya mill, a new ball mill and flotation plant were erected to treat the Mt Charlotte ore. With gold rising in price, the decision was made in 1979 to dewater many of the idle shafts. The Chaffers shaft was re-equipped and two of the most capacious shafts, Lake View and Perseverance, were 'rehabilitated'. The word was appropriate because a shaft and its winder house are like an extraordinarily strong and lanky giant who carries a host of men and supplies downwards, and all the mined ore upwards.

A burst of re-assessing and exploring took place in shallow and moderately deep ground. It is strange to think of successful exploration taking place within the accepted boundaries of a goldfield that was nearly 90 years old, but vast was the potential gold-bearing area. The Golden Mile itself was closer to a Golden Two and a Half In the translating of nearly all mine plans to the metric system in 1973, the Golden Mile was loosely defined as an area 4000 metres long by 1200 metres wide. As the gold-bearing ground extended about 1500 metres below the surface, the whole block of ground represented a huge exploration task when exploration first began in 1893. In sheer mass it was the equivalent of a buried corridor of ground stretching all the way from Melbourne to Adelaide, nearly one kilometre in width, and averaging some ten metres from top to bottom in depth. Even 85 years after the finding of the 'Four Golden Kilometres', it was still possible to discover, far below the surface of the leases held by the old mining companies, new lodes and extensions to old lodes. Moreover, under new methods of working, many areas half-mined in bygone days were seen once again as promising.

Mining on a fairly large scale was resumed below the Golden Mile itself in 1981. To mark the event, a ceremony was held at the new treatment plant built for the refractory Golden Mile ores as distinct from the friendlier Mt Charlotte ores. On Friday 10 July 1981 the Premier of Western Australia and great promoter of its mines, Sir Charles Court, visited Kalgoorlie to open the plant. Sir Laurence Brodie-Hall explained all that had happened, the obstacles that had been hurdled, the miracles in gold prices that had occurred. He thanked all the men who had managed the combined operations at Kalgoorlie in the last nine years — John Oliver, Jack Manners, Dick Hooker and Jack McDermott. The staff he thanked for their willingness even in leisure hours to write endless reports and revised reports and calculate the costs of each project and each deviation from a project. He praised the unions, especially the AWU, for their understanding of the peculiar economics of gold mining, for their willingness in the bad years to accept lower wages than those prevailing in other mines in order to keep the field alive until those recent gilded years when their wages more than caught up. There had been no major strike for 46 years, and many instances of close co-operation: 'Long may this spirit continue,' he said. He also drew attention to the role of chance, to the arrival of 'a fair share of good luck, divine providence or whatever'. In the next few years the hand of providence became more visible.

The price of gold was like a yo-yo but the average price was favourable in the early 1980s. The less attractive price for nickel was also a help, because the big nickel mines in the region ceased to compete for labour and so push up the wages in gold mining. Moreover a new mineral process, as advantageous as the Scottish cyanide process of the 1890s, helped gold mines; and the first carbon-in-pulp treatment plant at Kalgoorlie was opened in 1983 to treat half a million tonnes annually from an open cut on the old Golden Horseshoe lease. The surge of profits from gold mining enabled Kalgoorlie to invest in the latest equipment, and so less labour was needed. Only 700 were employed by Kalgoorlie Mining Associates by the middle of 1983, but it was again the biggest gold producer in Australia. That title was soon challenged, for the high gold prices set in motion the most vigorous gold boom since the 1930s. Far from the Golden Mile, mine after mine was being equipped for large-scale production. Even Kambalda, the nickel field, was now a major producer of gold.

Big-time promoters dabbled in gold mines for the first time since the 1930s. Alan Bond began to fly occasionally from Perth to Kalgoorlie, appearing in a grand promotion on Sunday 10 November 1985 at the new Mt Percy gold mine at the northern end of the field, two kilometres from Kalgoorlie. It was the kind of event rarely seen since the 1890s but now performed in jet-age style. The big party of guests was served champagne and croissants in a hasty breakfast at the Perth airport before flying to Kalgoorlie where they made a brief tour of the surface of the reborn North Kalgurli and the new Mt Percy. No expense was spared, and five hostesses 'dressed in raunchy gear' and each possessed of a 'wonderful personality' were there to escort the official guests. In an open marquee with seats for 300 they heard Alan Bond say that Mt Percy, which employed 53 and had poured its first gold the previous month, was an early sign of his 'long-term goal to become a major gold producer'. The manager, Brian Phillips, presented a painting to the Premier, Brian Burke, after his speech, and Bond gave a cheque to the senior citizens' community centre, and at 11.47 — according to the official program — the opening ceremony ended with a three-minute stunt.

In the stunt an old miner came towards the tent, pushing a wheelbarrow containing a bar of shining Mt Percy gold to display to the guests. Just as he called out the name 'Mr Bond', the gold robbers appeared. There was a brief scuffle, after which the master of ceremonies announced proudly that in the history of the goldfields an armed robbery had never been successful. So Mr Bond made his first impression on the goldfield he was to reshape.

Alan Bond was then aged 47. Born in London and educated in Fremantle, briefly a blue-collar worker, he had become one of the biggest investors and property developers in Australia. Without his money and drive, Australia would not have won the America's Cup in 1983. His interests soon were to stretch from Hong Kong to Chile, from a television network to a private university. Mining was already a major interest, but not gold. Bond's first big mining venture had been Metals Exploration Limited which owned half of the Greenvale cobalt and nickel mine in North Queensland, and a treatment plant near Townsville as well as various other mines including the Nepean nickel mine near Coolgardie. Through Metals Exploration, Bond became interested in the Golden Mile. In 1982 his company bought a 35 per cent interest in the North Kalgurli Mines, originally a London but now a Perth company. North Kalgurli had retained its important leases on the Golden Mile and by 1981 was again mining gold ores, treating them in the old mill that had temporarily been diverted to nickel production. It was North Kalgurli which, just before the opening of the Mt Percy gold mine in 1985, bought 43 per cent of the shares, thus enabling Alan Bond to preside in the day-long celebration.

The price of gold boomed, spurring Bond and his Metals Exploration to concentrate more on gold in 1986 and 1987. In a complex chain of transactions — which Bond sometimes clinched or decided upon while flying in his own jet — he gained a larger stake in the Golden Mile including the old Paringa mine. At the Golden Mile the revived company, North Kalgurli Mines, was his vehicle and chequebook. North Kalgurli was so busy buying up ground that its shares, worth a mere 45 cents each in June 1986, reached $1.34 in October. Higher gold prices fanned the excitement.

There was a pattern in Alan Bond's orgy of buying. He wished to control the whole Golden Mile, so that his companies could open a vast open cut or quarry that would engulf all the upper workings of all the old companies. With mechanised mining on a large scale he hoped to mine everything within a few hundred metres of the surface. He would remove it all — the totally barren ground, the tiny pockets of rich ground, and, above all, the spacious zones of very low-grade ores that had never been payable with the dearer underground methods. A huge open cut of that magnitude was possible only if Bond owned a continuous sweep of ground with no alien patches in the middle. He had to buy everything. That meant he would have to pay very high prices to the obstinate owners of little left-over leases that were not even capable of supporting a mine. He was prepared to

THE GOLDEN MILE

Hundreds of such scenes were photographed in the first years of the underground mines, and miners who kept the photographs were reminded years later of the flicker of the candles, the sound of an ore truck approaching in the darkness, the chattering noise of the rock drills, and the creaking of timber which supported the roof. In the 1990s these old working places were systematically excavated by the mining operations in the Super Pit. Today many such scenes are in mid-air.

pay for them. For a few years his gigantic cheque book floated above the Golden Mile.

The first major merger of scattered leases had taken place back in 1973 when the two main survivors, the Gold Mines of Kalgoorlie and the Lake View & Star, merged their operations and consolidated their leases. The child of that consolidation, Kalgoorlie Mining Associates, therefore had to be approached if Bond's expanding ambitions were to be achieved. Either they had to buy him out — an unthinkable idea for Bond — or he had to buy them out. He decided to buy, fortified by his excessive faith that the price of gold would remain high. As Kalgoorlie Mining Associates was not a listed company, it could be acquired only indirectly. Moreover, it consisted or a mere two shareholders, one of which was the American firm of Homestake — apparently not a seller. Accordingly the only path of access was through the other shareholder, Kalgoorlie Lake View, which held the 52 per cent interest.

In 1987, a bubbling year on the world's stock exchanges, Alan Bond began to bid for control of Kalgoorlie Lake View. It was largely owned by two main mining groups, Poseidon of Adelaide, and the Melbourne company, Gold Mines of Kalgoorlie. Bond first turned his chequebooks onto Gold Mines of Kalgoorlie and for a high price he gained more than half of the shares. That gave him effective control of 47 per cent of his secondary target, Kalgoorlie Lake View (which in turn owned 52 per cent of his principal target, Kalgoorlie Mining Associates). It was crucial for Bond to increase his 47 per cent to 53 per cent. He thought he saw a simple solution. He could buy out Western Mining's small but decisive share of 6 per cent in Kalgoorlie Lake View, thus giving him control of nearly all of the Golden Mile.

At this stage Bond was thwarted. Apparently he did not realise that Poseidon Limited had a pre-emptive right to buy half of the Western Mining interest of 6 per cent. Poseidon promptly bought its 3 per cent from Western Mining; Bond bought the other 3 percent. So there was a deadlock. The key company, Kalgoorlie Lake View, was thus owned 50-50 by Bond and Poseidon. Bond, for all his strenuous buying, still did not control the field.

So Western Mining sold out of the field that had been for so long a key part of its history. No other company had done so much to keep Kalgoorlie alive in the 1950s and 1960s and 1970s. Western Mining sold out partly because it now had many interests of far greater importance than Kalgoorlie. A world leader in nickel, Australia's largest producer of gold, the owner of 51 per cent of the shares in the huge copper-uranium deposit at Olympic Dam in South Australia, and the main Australian shareholder in Alcoa Australia which led the world in alumina, Western Mining had enough on its hands. Reluctantly Western Mining accepted that its relatively small stake in Kalgoorlie now merited a low priority. It was also persuaded to sell out of Kalgoorlie by the high prices offered. Furthermore it had been deploying more technical and managerial talent in Kalgoorlie than was probably justified by its small financial stake in that field. Now Western Mining's only stake there consisted of leases of the old Great Boulder, which it had fully owned since 1976, and the nickel smelter it operated to the south of the Golden Mile.

Eventually the deadlock between Bond and Poseidon was resolved, at a high price. Allister McLeod, the outgoing chairman of Poseidon and Robert de Crespigny, the incoming chairman, agreed that the 'orderly and timely development' of the Golden Mile required one owner rather than divided control. At the end of 1987, Poseidon sold out its 50 per cent interest in Kalgoorlie Lake View to the Bond group. The price was generous — it had to be — and unusual: $200 million in cash, $25 million in notes that could be converted into Gold Mines of Kalgoorlie shares, and a total of 214 000 ounces of gold currently worth about $150 million. The gold was to be paid from production over the following six years, and GMK's assets acted as security for this continuing payment.

Barely two years after the ceremony in Mt Percy's marquee at the downmarket end of Kalgoorlie, Bond had almost achieved his ambition. He was chairman of his private Dallhold Investments Pty Ltd and the three relevant public companies — Metals Exploration, North Kalgurli Mines, and Gold Mines of Kalgoorlie. It was an intricate chain of command. His Dallhold Investments owned 58 per cent of Mideast Minerals No Liability; which in turn owned 98 per cent of Metals Exploration, the big nickel company; which in turn owned 38 per cent of North Kalgurli Mines; which in turn owned 51.5 per cent of Gold Mines of Kalgoorlie; which, in turn, after the Poseidon deal, owned 52 per cent of Kalgoorlie Mining Associates. The other partner in Kalgoorlie Mining Associates was Homestake which owned 48 per cent of the shares.

Perth, through Alan Bond, at last controlled the goldfield which for more than half a century had been largely directed from London but more recently from Melbourne and Adelaide. For the first time one businessman and his tangle of interests — and debts — controlled nearly every key lease inside the Golden Mile as well as the producing mines of Mt Charlotte and Mt Percy at the Kalgoorlie end of the field. Bond won command of the Golden Mile only because he was willing to offer very high prices — he was a skilled raiser and borrower of funds.

Alan Bond announced bold plans to expand Kalgoorlie by turning the upper zone of the Golden Mile into a huge open pit. On the strength of the reborn Kalgoorlie operations, he said he would build up a gold-mining house to rival the mighty South African gold houses. While he was not the originator of the idea of one big operation, he did more than anybody to make it possible. It was probably the most constructive act in his business career but the money he paid for the parts of his jigsaw was too high. The price of gold did not stay as high as he had hoped. Many of his other commercial ventures were struggling. In

August 1989, feeling the financial pressure, both he and his bankers were looking for a company that would buy out the Bond interests in Kalgoorlie. Suddenly the Adelaide company, Poseidon, was back in the running again.

Robert Champion de Crespigny of Poseidon was about to become the key individual on the Kalgoorlie field. His mining business had grown almost from nothing. A commerce student at Melbourne University, he had gone to Perth, becoming a chartered accountant with interests in the resource companies that flourished in the series of booms in Western Australia. In December 1985 he acquired 20 per cent interest in a small Perth company which became Normandy Resources NL. (The de Crespigny family, at the time of France's persecution of the Huguenots, lived in Normandy.) Under the new management, Normandy Resources expanded quickly. Primarily a 'resource finance company', it invested in a variety of mineral ventures ranging from oil-searching in the Dutch North Sea to gold mining in many parts of Australia. In 1987 it had entered the bigger mining league with its purchase, from the Anglo American Corporation of South Africa, of a stake of 18 per cent in Poseidon, through which it indirectly acquired an interest in the Golden Mile. Normandy was now the largest shareholder in Poseidon, and Robert de Crespigny soon became chairman of Poseidon. Just before he became chairman it was he who suggested that Poseidon use its right to pick up another 3 per cent in Kalgoorlie Lake View, thus enhancing Poseidon's bargaining position when it decided to sell out its 50 per cent to Bond.

Poseidon, like nearly all those who sold gold leases and shares to Bond, did well out of the 1987 sellout. The company was also in a powerful position, should the occasion arise, to buy back what it sold. If Bond one day was to walk out of the Golden Mile — in other words sell out his shares in the crucial Gold Mines of Kalgoorlie — his only likely purchaser was Poseidon, because it had a security over GMK's assets. Less that two years later, financial buffeting made Bond eager to sell. On the other hand de Crespigny, as chairman of Poseidon, was prepared to buy into the Golden Mile only if the price was right. Naturally Bond's selling price was much lower than his price of two years ago: he urgently needed the money. In a complicated deal Poseidon bought 19 per cent of the shares in Bond's Gold Mines of Kalgoorlie with the option to increase that holding to 29 per cent. Becoming the main shareholder in Gold Mines of Kalgoorlie, Poseidon also won the right to manage the Golden Mile. And so the control of the Golden Mile passed to Poseidon or, as it was called from 1991, Normandy Poseidon. In effect Poseidon had merged with its largest shareholder, Normandy Resources, to form the new Normandy Poseidon.

Already the big pit or open cut was firmly on the drawing board. Some 8000 holes had been drilled and well over 100 000 assays made in order to determine that the big low-grade gold mine would be profitable. In recent years a variety of smallish open cuts had been excavated along the Golden Mile, and now they were to be successively absorbed into one big quarry that would take out nearly all the Golden Mile at shallow depth. Eventually the open cut, it was hoped, would extend down 500 metres — if the price of gold was not overtaken by a renewed inflation of mining and living costs. According to estimates made in 1991, about 90 million tonnes of ore would be mined on the grand scale and sent in heavy trucks to treatment plants. In addition, hundreds of millions of tonnes of barren or unpayable rock either overlaying or separating the twenty main lodes would be quarried and sent to the dumps.

The venture was based on mining the rock which the old companies, in their early underground activities, either had missed or rightly had decided was not payable. The success of the venture depended on ore carrying little more than two grams to the tonne, of which nine grains of gold in every ten were extracted in the treatment plant. Converted into old measurements, the ore carried less than two pennyweights to the ton compared to more than 20 pennyweights in the average ore of the typical mine during the early years of this century. Ian Burston, the chief executive at Kalgoorlie, pointed out in 1991 that the mining of such ore even a decade ago 'would have seemed ludicrous to contemplate'. Once famous for its richness, Kalgoorlie's ore was now best known for its poorness. The goldfield's survival, on its hundred birthday, owed far more to the ingenuity of miners than to the generosity of nature.

Much of the mining history of Kalgoorlie was the story of each new generation of miners, improving on the mining, hauling and ore-treating methods of their fathers, and so winning a profit from gold-bearing rock which previously was unpayable. The super pit was a dramatic example of using new techniques to rework old areas, blasting big blocks of ground, using gigantic shovels to load the 130-tonne trucks which hauled the ore to treatment plants working on a mammoth scale. With the new methods, one man could mine and handle as much ore as 30 men mined at the start of the century.

The project would have been impossible without another innovation, the carbon-in-pulp process which cheaply recovered a high percentage of the gold from oxidised ore and the simpler sulphide ore. The process relied on crushing and grinding the ore, the use of cyanide to dissolve the gold into solution and carbon to catch the gold in solution. Equally important was the use of the mechanical backhoe which enabled the pockets of payable ore to be mined in the pit with some precision, leaving the surrounding barren rock to be mined separately and carted to the waste dumps.

To treat the increasing tonnage of ore the new, united Kalgoorlie employed four treatment plants near the big pit: the new Fimiston opened in 1989 and doubled two years later, and the Oroya, Croesus and Paringa plants. Further north the Mt Percy plant treated a large tonnage of low-grade ore — by 1991 it had the remarkable safety record of suffering not one lost-time accident in nearly six years. The fifth treatment plant lay 17 kilometres along the Great Kalgoorlie-Leonora road, at Gidgi. It roasted the concentrates, high in gold, that came from the four treatment plants near the big pit.

On the eve of its centenary, Kalgoorlie consisted of four separate mining operations. In the far south, near Boulder, was the last of the underground operations on the true Golden Mile; then came the big open cut now called the Super Pit; further north the poppet head of the largest underground gold mine in Australia looked down from Mt Charlotte onto the main street of Kalgoorlie; and finally could be seen the waste dumps of the small Mt Percy open cut. About seven tenths of all the gold was coming from very low-grade ore in the Super Pit, now the largest gold mine in Australia. The field itself was again the largest in Australia, producing in 1991–92 more than 600 000 ounces of gold, or a tally higher than in any year since the first decade of the twentieth century.

So Kalgoorlie celebrates its centenary by mining the relics of its past. In the space of a few years the explosives and great long-necked mechanical shovels and huge trucks — heavier than are normally allowed on a highway — have removed enough ore to form this vast pit, delving deeper every month. Most of the poppet heads and the clusters of mine buildings on the old Golden Mile have already gone — nearly all that remain will soon disappear.

The destruction of landmarks drew small crowds of spectators. It was a memorable day when the celebrated Perseverance poppet head, originally erected on the Great Fingall mine in the Murchison, was to be blown up to make way for the fast-advancing open cut. The charges, carefully prepared, were fired. Some spectators said they felt lumps in their throat as they thought of all their relatives and friends who each working day had assembled under that stately poppet head for the long journey down to their workplaces. But seconds after the firing, when the smoke and dust had cleared away, they were astonished to see the three thick steel legs still standing firm in their foundations of concrete. In the end the legs were amputated low down, and the poppet head topped, almost intact. The scene reminded one spectator of a dead elephant lying on its side, on the edge of a quarry.

Much of the rock shallow once supporting small towns to the east of the road between Kalgoorlie and Boulder has been carted away. Part of the area traversed by the busy trams and the railway loop before the First World War is now in thin air. Women returning after an absence of 60 years from the goldfields look down into the pit, and puzzled by the loss of landmarks, ask where is the site of the school and Sunday school they attended as children, the shop where they bought boiled lollies, the pepper tree which shaded them on hot days, the hotel where their father sometimes drank. So often, in answer to their ques-

THE GOLDEN MILE

The 'celebrated Perseverance poppet head', standing like a Meccano exhibit in 1979, was toppled to make way for the advancing pit.

170

tions, their eyes are directed to the big hole in the ground or to the encircling hills of new-mined waste rock.

Sometimes on the walls of the pit can be seen the old underground tunnels briefly exposed for the first time to daylight. Here long ago, in the candle-light, miners ate their crib, told their yarns, reminisced of Moonta and Maryborough, did their hard day's work and even died. Their workplaces of 1899 and 1909 are briefly brought to light by heavy charges of explosives and then rapidly mined. Occasionally the nail of a miner's boot or a tobacco tin or a clay pipe, relics of perhaps 90 years ago, find their way with the ore to the huge gold-extraction plants which now crush more rock each day than ever before.

The gold at Kalgoorlie is so sparse that it is little more than a vapour of gold. As in the foundation year of the Golden Mile, most newcomers still exclaim, after examining freshly-mined rock, almost the same thoughts as were voiced in the bough shed at the baby Great Boulder to Sam Pearce and Will Brookman in 1893: 'But where's the gold?' Rarely visible in the rock, the gold of Kalgoorlie was almost as vital in the early 1990s as in previous nation-wide depressions. After its long sleep, gold was again one of Australia's three major exports, and Kalgoorlie was the biggest producer. Never in the history of Australia had a mining field shown such resilience, such a gift of rising again from the dead.

THE GOLDEN MILE

By the end of the 1992 the 'Super Pit' at Kalgoorlie was eating deep into the old underground workings and the bordering areas of unmined rock.

Loading one of the heavy trucks in the Super Pit. Every ten tonnes of average ore yielded, after treatment, just under an ounce of gold.

The edges of the Super Pit, expanding every month in 1992, were approaching the poppet heads of mines, the rusting machinery that was once an engineer's pride, and the shrinking oases of greenery that marked the site of offices and houses and shops.

SOURCES OF INFORMATION

THE BOOK RELIES much on information published in a variety of technical reports on the Kalgoorlie goldfield. These included annual reports of the Department of Mines, beginning in the 1890s and usually providing statistical as well as technical detail, especially by the inspectors of mines and, later, inspectors of machinery. For the dismal 1920s the reports of the state mining engineer were useful.

Statistics

Figures on gold output, population and other changes are in successive editions of the *Western Australian Year Book*. Dividends paid by companies in early years were published regularly by the Chamber of Mines based in Kalgoorlie. Several of the statistical comparisons appearing in this book — for example the comparing of Western Australian gold dividends with Australian bank dividends in 1907 — are the result of rough calculations using available statistics. A few comparisons — for example the relative capital cost of railways and the water scheme in Western Australia — are based on historical statistics from such standard sources as T. A. Coghlan, *A Statistical Account of the Seven Colonies of Australasia* (Sydney, 1898).

Individual mining companies

The Chamber of Mines published a *Monthly Report* from 1903 and a *Monthly Journal* from 1906. They hold much of value on mining conditions, gold stealing, and technical changes. For the plight of the goldfield in the 1970s the reports of the Chamber's annual meetings were especially useful. Annual reports of some of the major old mining companies are not easily found. Especially useful were annual reports and plans, and verbatim accounts of shareholders' meetings, of Great Boulder Proprietary from 1895 to 1908: they are held by the Chamber of Mines in Perth. Useful reports and summaries of individual gold companies and their dividends are to be found in scattered year books including the long-running *Mining Year Book* edited in London by Walter E. Skinner; R. L. Nash, *Australasian Joint Stock Companies' Year Book* (Sydney, 1913–14); and the annual *Mining Handbook of Australia*, prepared by staff of the *Chemical Engineering and Mining Review*, and published in Melbourne from 1936. In London in 1948 the *Mining World and Engineering Record* was publisher of an anonymous book, *Westralia: Story of the Goldfields*, giving much space to recent changes in the big Kalgoorlie mines. The printed annual reports of modern mining companies are valuable, and I thank Brian Phillips for access to annual reports of the Bond companies and Nicholas Smith for information on Normandy Poseidon.

Little has been published on the history of the many mining companies which flourished at Kalgoorlie. To my knowledge there is nothing on such a notable mine as Great Boulder. Another famous mine is documented in a company booklet, *Fifty Historical Years: The Story of Lake View and Star Limited*, covering the years 1910 to 1960. On a relative newcomer, Gold Mines of Kalgoorlie, there are short sections in the book by Sir Lindesay Clark listed below. In September 1984 Gilbert M. Ralph completed 'A Brief History of Gold Mines of Kalgoorlie Limited', published in ten pages.

Parliamentary papers and reports

In the Western Australian parliamentary papers are C.Y. O'Connor's 'Coolgardie Goldfields, Report on Proposed Water Supply (by Pumping)', 1896; the illuminating royal commission into the Great Boulder Perseverance Gold Mining Co., 1905; the royal commission 'on pulmonary diseases amongst miners', 1910–11; and the Kingsley Thomas royal commission that reported in 1925 on Kalgoorlie's mines and their isolationism. The W.A. Hansard reported, beginning on 29 August 1922, the debate on the bill introduced by J. Scaddan to combat miners' disease.

The Commonwealth of Australia published in 1901–02 a report on 'Foreign Contract Labour' in W.A., especially the Italians. Another useful report is H.A. Hunt, *Results of Rainfall Observations made in Western Australia* (Melbourne, 1929). It prints the early goldfields rainfall records which O'Connor omitted from his reports. In the Commonwealth of Australia Hansard is the House of Representatives debate on the Gold Bounty Bill on 10–11 December 1930, and the debate on the Gold-Mining Industry Assistance Bill, 4 and 10 November 1954.

Technical reports and books

Among those entirely or partly on Kalgoorlie are Donald Clark, *Australian Mining & Metallurgy* (Melbourne, 1904) with its first 116 pages on Kalgoorlie mines; Ralph S.G. Stokes, *Mines and Minerals of the British Empire* (London, 1908); K.S. Blaskett, *Ore Treatment at Western Australian Gold Mines* (Melbourne, 1952); Rex T. Prider (ed.), *Mining in Western Australia* (Perth, 1979); A.L. Lougheed, *The Cyanide Process and Gold Extraction in Australasia, 1888–1913* (Economics Dept, University of Queensland, 1985); A.B. Edwards, *Geology of Australian Ore Deposits* (Melbourne, 1953); and E. Davenport Cleland, *West Australian Mining Practice* (Kalgoorlie, 1911).

Other books, pre-1914

On the period before 1914 I gained much from Harry P. Woodward, *Mining Handbook to the Colony of Western Australia* (Perth, 1895); Julius M. Price, *The Land of Gold: The Narrative of a Journey through the West Australian Goldfields in the Autumn of 1895* (London, 2nd edn, 1896); May Vivienne, *Travels in Western Australia* (London, 1902); Arthur Reid, *Those Were The Days* (Perth, 1933), for which Keith Quartermaine of Kalgoorlie has lately compiled a useful index; G.H. Nash, *The Life of Herbert Hoover* (New York, 1983); W.G. Manners, *'So I Headed West'* (Kalgoorlie, 1922); and Ian Hore-Lacy, *Broken Hill to Mount Isa: The Mining Odyssey of W.H. Corbould* (Melbourne, 1981). On several facets of the social history of Kalgoorlie-Boulder at the turn of the century I gained much from Norma King, *Daughters of Midas: Pioneer Women of the Western Goldfields* (Victoria Park, W.A., 1988) and Andrew Rick-

SOURCES OF INFORMATION

etts, *Walter Lindrum: Billiards Phenomenon* (Canberra, 1982).

On the role of Pearce and the Brookmans I am indebted to R. M. Gibbs, *Bulls, Bears and Wildcats: A Centenary History of the Stock Exchange of Adelaide* (Adelaide, 1988) and to his article 'The Real Poseidon' in the *Journal of the Historical Society of South Australia*, no. 4, 1978. Henry Lawson's description of Modeste Maryanski is in his 'The New Westward Ho!' reprinted in Colin Roderick ed., *Henry Lawson: Autobiographical and Other Writings 1887–1922* (Sydney, 1972) pp. 73–8. My essay on Patrick Hannan can be found in Lyall Hunt ed., *Westralian Portraits* (Perth, 1979) pp. 73–7. Some of the important documents and reports on water are printed in full in F. Alexander, F. K. Crowley and J. D. Legge, *The Origins of the Eastern Goldfields Water Scheme in Western Australia* (Nedlands, 1954).

Books on the period 1914–60

Charles A. Price, *Southern Europeans in Australia* (Melbourne, 1963); L. R. M. Hunter, *Woodline* (Forrestfield W.A., 1976); Danny Sheehan with Wayne Lamotte, *Heads and Tails* (Kalgoorlie, 1985); and G. Lindesay Clark, *Built on Gold: Recollections of Western Mining* (Melbourne, 1983). A volume of short stories, edited by Ted Mayman, *View from Kalgoorlie* (Perth, 1969) is revealing about working life in the mines, especially the stories by Gavin Casey.

Archives

On the water crisis of 1893 and 1894 the Western Australian State Archives have useful correspondence from Finnerty (especially File ACC 964). On some episodes — for instance the diversity of times kept on the Golden Mile in April 1902 — correspondence is held by the Chamber of Mines in Perth. Most of the Chamber's early correspondence is with the W.A. State Archives where I consulted the papers 'Foreigners in Mines' for the war years 1914–18 and 1940. I gained much from the long, unpublished *An Autobiography* by the Boulder miner, S. R. Whitford, held in the Mortlock Library of South Australiana.

Newspapers and journals

Often the small detail about events comes from the daily *Kalgoorlie Miner*, first published in 1895. For example on 11 February 1928 it reported the 'Disastrous Cyclonic Storm'. The *Ballarat Courier* from July 1895 to June 1896 was read for news of people departing for Western Australian goldfields; it also contained advertisements for passenger ships about to sail for Albany. The opening of the water scheme at Mundaring, Coolgardie and Kalgoorlie is described in the *West Australian*, 19, 22, 23, 26, 28 January 1903. Detail about mining operations is sometimes from the *Australian Mining Standard*, especially the years 1895, early 1897 (on tellurides and Harry Rhys Jones's predictions about low-grade ore), 1910 and 1922. The *Clarion*, a lavish journal edited and almost entirely written by Randolph Bedford, devoted its issue of February 1897 to the Western Australian goldfields. The Melbourne

Age sent 'our special reporter' to W.A. and two of his long articles on Kalgoorlie were published on 18 and 25 July 1896. At Kalgoorlie the Chamber of Mines has news-clipping books with information on working conditions which the state's Arbitration Court was investigating in 1921. Other clippings tell of the managers' reactions to Kingsley Thomas's criticisms in 1925.

Photographs

Nearly all the early photographs come from the Museum of the Goldfields in Kalgoorlie. Most were taken by J. Dwyer, a skilled local photographer, whose glass plates were acquired by the museum. The photos on pages 4–110 (and also page 164) are from this collection, with the exception of the items listed below.

The photo on page 24 was provided by Mr Philip Lovely of Hawthorne, Queensland.

The photos on pages 28, 36 and 64 are from the J.S. Battye Library of West Australian History in Perth.

The photo on page 85 is from Great Boulder Proprietary's annual report for 1906.

The photo on page 90 is from the Chamber of Mines and Energy of W.A. in Perth.

The photos on pages 102 and 103 are from the Eastern Goldfields Historical Society in Kalgoorlie.

The photo on page 142 is from West Australian Newspapers Limited.

Of the post-1930 photographs (page 125 onwards), most are from Western Mining Corporation's collection in Melbourne. Several of the most recent photographs are from Kalgoorlie Consolidated Gold Mines, Kalgoorlie, and were gathered by Norma Latchford. Three of the photographs belong to the author.

INDEX

Adelaide (SA), mining interest, and Kalgoorlie field 6, 8, 9, 16, 18, 19, 37, 48, 166
Adelaide Steamship Co. 23
Age, Melbourne, journalist 29
Agnew, John 129, 130, 150
Albany (WA) 7, 23, 34
Alcoa Australia, and alumina 165
All Nations Boarding House, Kalgoorlie 141
amalgamating tables, pans and processes 18, 46
Anderson, Ralph 131
Anglo American Corporation of South Africa 167
aqueducts, in Italy and England 59
Arbitration Court, and minimum wage 114
'Ardmona', (Richard Hamilton's house) 89
Associated Gold Mines of WA: company and mine 16, 35, 37, 115, 130; dump 88; mine shaft 84; strike (1916) 107
Associated mines, company offices 88
Associated Northern Blocks, company and mine 88, 89, 100
Aswan dam project, Egypt 65
Australasia Chambers, Adelaide (SA) 8
Australasian Institute of Mining and Metallurgy 118
Australia Hotel, Kalgoorlie 54
Australian Mining and Metallurgy 12
Australian Workers' Union (AWU) 113, 118, 143, 144, 161
Austrians, on goldfields 106

backhoe, mechanical 168
Ballarat (Vic.) 1, 23, 35, 48, 50, 60, 92, 103; School of Mines 44, 134; South Street Society 94
balls, steel, for Krupp mills 108
Bank of Ireland Gold Mining Co. 32
basalt 12
batteries, mine 19
Bavarian Band 63, 65
Bayley, Arthur 1
Bedford, Sir Frederick 50
Bedford, Randolph 19, 21, 25
Bendigo (Vic.) and goldfield 23, 29, 35, 40, 42, 46, 50, 84, 89, 139, 150
Bernales, Claude A. de 134–5
Bewick, Moreing and Co. 91, 129, 151
Big Bell mine 139, 147
billiards 54
Black, R.S. 89
Blackett, Charles 118
Blainey, Sam 25
Block 45; mine 32
blowers, mechanical 118
Blue Spec, racehorse and mines 50
'bogger', mechanical loader 145
Bond, Alan 162–3, 165–7; plan for open-cut mine 163, 165, 166
Boulder Bonanza mine 30
Boulder Central Extended mine 29
Boulder Central mine 29
Boulder [City]: brass bands 94; Croatian Hall 106; cyclone (1928) 126; decline in 1920s 114, 126; early appearance 47, 48, 50; managers' houses 88; mines 30; municipal council 107; pipeline opening 66; railways to 103; recovery in 1930s 128, 139; reef 30; riot (1934) 141; shopkeepers 91; town life staider 92

Boulder Half-mile South mine 30
Boulder Main Reef mine 88
Boulder North Extended mine 29
Boulder Orchestral Society 50
Boulder Perseverance mine; see Perseverance
Boulder Racing Club 82
Brinsden, F.G. 134
Brisbane (Qld) 48
Broad Arrow (WA) 54; road 135
Brodie-Hall, L.C. 157, 159, 161
Broken Hill (NSW) field 11, 14, 37, 44, 61, 70, 100, 117, 119
Broken Hill Proprietary Co. (BHP) 134, 152
Brookman, Benjamin (sen.) 8
Brookman Bros Boulder Co. mine 29–30
Brookman, George 6, 7, 8, 13–14, 19, 37; original syndicate 6–8; raising capital 14; *see also* Coolgardie Gold Mining and Prospecting Co. WA Ltd
Brookman, William G. 6, 7–8, 9, 11–12, 13, 37–8, 171
Brownhill mine 100, 124; *see also* Hannan's Brownhill
Bruce, S.M. 117
Buckett, Reg 148, 150
Bull, Eli 38
Bulong (WA) 32, 76
Burke, Brian 162
Burston, Ian 168

calaverite (mineral) 34
calc schist rock 80, 115
Callahan, Henry 88
camels and camel teams 5, 9, 22, 47, 65
Cameron, Keith 151

Carroll, John (Jack) 103, 106
Casey, Gavin 136
Cassidy shaft, Mt Charlotte mine 161
cement 107, 144
Central and West Boulder mine 29
Central Norseman mine 150
Chaffers mine 80, 161; treatment plant 148
chalcite rock 32
Chamber of Mines, at Kalgoorlie 79, 83, 107, 113, 115, 122, 143; moves to Perth 157
Champagne Syndicate NL 151n
Charters Towers (Qld) 6, 32, 35, 61
Clarion 19, 25
Clark, Donald 12, 42
Clark, Sir G. Lindesay Clark 145, 151–2
classifier, in crushing 46
climate *see* Kalgoorlie: climate and rainfall
Clunes (Vic.) 23
coal 103, 150; coal tar 117; mine (NSW) 119
codes, mining 89
coke, Welsh 107
Collier, Philip 97, 119, 141
commissions, royal 91, 96, 97, 107, 116
communists, in Kalgoorlie 119
compressors, for air 77, 124, 143
condensing, machines and practice 5, 58, 61, 71, 74, 103
Congregationalists, in Adelaide 8
Conolly, H.J.C. 152
Consolidated Gold Fields of South Africa 129, 150
contract mining 96
Coolgardie (WA) and goldfield 1, 2, 5, 7, 8, 22, 27, 91; and goldfields water scheme 56, 59, 60, 61, 62, 63, 65, 69, 76; rainfall 69

Coolgardie Gold Mining and Prospecting Co. WA Ltd 8, 13, 37; *see also* Brookman, George: original syndicate
Coolgardie Goldfields Universal Water Supply Scheme 61
Coolgardie Miner 58
Coral gum (Eucalyptus sp.) 13
Corbould, W.H. 3, 11
Cornishmen, on goldfields 92
Cottesloe (Perth) 134
Country Party 119
Courier, Ballarat 23
Court, Sir Charles 161
Cracow (Qld) 32
Crespigny, Robert Champion de 166, 167
'crib' (lunchtime break) 135, 136
'Crimson Flash' *see* Postle, Arthur
Cripple Creek, Colorado (US) 42, 88
Criterion Hotel, White Feather 56
Croatians, on goldfields 106, 107
Croesus (Proprietary) mine 34, 38; treatment plant 168
Crossley, Ada 16
crucibles, in assay shops 107
crushing machines 18, 79; *see also* stampers, mine and mill
Cumpston, J.H.L. 97
cyanide process (MacArthur-Forrest) and cyanide 40, 42, 44, 46, 79, 86, 136, 162, 168
cyclone (1928) 126
Cygnet lease 82

Dallhold Investments Pty Ltd 166
Dalmatians 106, 107

Darbyshire, J. 107
Dashwood's Gully (SA) 6, 7, 9
Day, R.B. 50
de Bernales, Claude A. *see* Bernales, Claude A. de
deaths, in mines *see* mines, and mining companies in Kalgoorlie: accidents and safety in
Defence Department, in war of 1914–18 112
Dehne press 46; *see also* filter press *and* Diehl, Dr Ludwig
Deland's Bakery, Kalgoorlie 30
depressions: of 1893 3; of 1930s ch. 8; of early 1990s 171
devaluation: of Australian pound 128; of English pound 144; of Australian dollar 160
Diehl, Dr Ludwig, and Diehl process 44, 46
diesel: power 130, 150; vehicles 156
dolerite 11, 12
dolly pot, and dollying 7, 13
Donnellan, Mr, of White Feather 56, 58
Doolette, (Sir) George 8, 14, 16, 115
Dower, Joseph 38
drills: hand 8; rock 77, 94, 118, 130, 131, 143, 145; diamond 25, 160
dry-blowing 3, 5, 7
dry-crushing, of gold 46; *see also* stampers, mine and mill
Duck Pond *see* Lake View Consols
Duke, Albert 124
dumps, mine (tailings) 79, 86, 88, 92, 136
Dunedin school of mines 129
dust, at mines 79, 88; *see also* mines and mining companies: dust and ventilation
dynamite 8, 38

INDEX

Eaglehawk (Vic.) 89, 97
East Coolgardie, name 27
Edwards roasting furnace 46
Egerton (Vic.) 23
Eiffel tower, Paris 84
Elvey, Edgar 147, 150
English language, use of 103, 141, 143
Esperance (WA) 61
Espie, Frank 151
eucalyptus 13, 117
Europeans, southern, and European immigrants 101, 103, 139, 143; see also Dalmatians; Italians, on goldfields; 'Slavs', on goldfields
Evening Star, Boulder 48
explosives 8, 38, 107, 131, 144, 145, 169, 171

Fanny's Cove (WA) 61
Feldtmann, W.R. 89
Ferguson, Mephan, Melbourne foundry 63
filter press 44, 46, 47, 74, 79; Dehne press 46
Fimiston treatment plant 168
Finnerty, John M. 5
fire: at Coolgardie (1895) 27; in Kalgoorlie mines 94
firewood 12, 19, 60, 71, 79, 101, 103, 108, 112, 116, 150; see also woodcutters and woodlines
Flanagan, Tom (prospector) 2, 3
flotation process and plant 117–18, 130, 135, 148
Ford, William 1
Forrest, Lady Margaret 65

Forrest, Sir John, and water scheme 55, 56, 58, 61, 62, 65, 66, 68, 75
Freemasons 92
Fremantle (WA) 23, 55, 60, 101
funds, shilling (health) 94; see also Mine Workers' Relief Fund

Gates Crusher (stamp mill) 46
Gawler (SA) machinery 19
gelignite 131
Geraldton (WA) 101
Germans, on goldfields 106
Gibbs, R.M. 8
Gidgi treatment plant 169
Gold Bounty Bill (1930) 128
Gold Exploration and Finance Co. of Australia 151n
Gold Mines of Kalgoorlie (GMK) 134, 143, 151, 152, 156, 158, 160, 165, 166, 167
Gold Mining Industry Assistance Bill (1954) 144
gold: nuggets 3, 50, 129; occurrence on Kalgoorlie field 2, 3, 8, 9, 11–12, 18, 30, 35, 130; output of gold and ore from Kalgoorlie field 35, 38, 40, 75, 77, 82, 83, 86, 92, 97, 100, 107, 122, 136, 139, 143, 147, 150, 162, 168, 169, 171; oxidised ore 40, 168; price 79, 107, 113–14, 128, 144, 145, 156, 158, ch. 10 *passim*; rushes 1–2; separation and treatment, and treatment plants 19, 40, 42, 44, 46, 74, 75, 79, 80, 96, 117–18, 130, 136, 148, 157, 161, 163, 168–9, 171; smelting rooms 46; stealing of 19, 82–3, 119, 128, 144; subsidy (bounties) 134, 144–5, 152, 159; sulphide ore and zone 12, 40, 168; see also tellurium and telluride ore
Golden Eagle nugget 129
Golden Gate company, Charters Towers (Qld) 61
Golden Gate railway station 79
Golden Horseshoe company and mine 77, 80, 84, 86, 92, 114, 118, 130, 162
Golden Horseshoe Estates 88, 122
Golden Link leases 38
Golden Mile Dolerite 12
'Golden Mile', name 50, 161
Golden Pebble lease 16
Golden Pike Fault 152
Goldfields Courier, Coolgardie 19
Gordon (Vic.) 23
government, federal, assistance to mines 128, 144, 159; see also gold: subsidy (bounties)
Great Australian Bight 1, 23
Great Boulder: discovery and name 9; early mine (Great Boulder Proprietary Gold mines) 11, 14, 16, 18, 19, 22, 29, 30, 34–5, 37, 38, 48, 50, 63, 77, 80, 88, 89, 91, 171; ore treatment at 42, 46, 86; shafts 83–4; site (1907) 77; after war of 1914–18 114, 115, 122, 134, 135, 139; in decline from 1960 147, 150, 156, 159; leases in 1980s 166; nickel treatment 156; see also Hamilton, Richard
Great Boulder Hotel, Kalgoorlie 54
Great Boulder Main Reef mine 101
Great Boulder No. 1 company 88
Great Boulder Perseverance *see* Perseverance
Great Boulder South mine 30
Great Fingall mine, Murchison 101, 169

Greeks, on goldfields 141
greenstone 11, 12
Greenvale cobalt and nickel mine (Qld) 163
Griffin mill 46
Grundt, William 135
Gum Creek, north of Menzies (WA) 61
Gympie (Qld) 32

Hackett, Dr J.W. 65
Hainault Gold mine 89; company's Glasgow office 88
Half Way, the 128
Hall, Alfred 83
Hamilton, Mrs Kate M. 89
Hamilton, Richard 42, 80, 89, 108, 117, 122, 126, 150
Hannan Street, Kalgoorlie 48, 112, 135, 141, 144, 151, 152
Hannan's [Find] 2, 3, 5, 7, 8, 13, 14; *see also* Kalgoorlie
Hannan's Brownhill company and mine 30, 35, 40, 44, 80, 89, 91; *see also* Brownhill mine
Hannan's Club, Kalgoorlie 83, 97
Hannan's Golden Morning Star mine 30
Hannan's Golden Pebbles shaft 27
Hannan's Golden Trees mine 30
Hannan's Lake 19
Hannan's North mine 134, 136, 152
Hannan's Oroya mine 80
Hannan's Reward lease 152
Hannan's Star company and mine 30, 38
Hannan's Star Consolidated company 100
Hannan, Patrick (Paddy) 1–3, 5, 6, 7, 9, 11, 12, 31, 48, 50, 152; death 124

Harper, Nat 58
Hartman, T. F. 88
health, of mine workers 94, 96–7, 115–16, 131; *see also* mines: accidents and safety in *and* mines: dust and ventilation
Hehir, Jack 135
Hewitson, Thomas 89
Hocking, miner at Great Boulder 114
Hogan, E.J. 103
Holroyd, Arthur 32, 34
Home From Home Hotel, Kalgoorlie 144
Homestake Mining Co. 159–60, 165, 166
Honali mine, India 89
Hooker, Dick 161
Hoover, Herbert 44, 91–2, 126, 127, 129, 143
horses: horseracing and racecourses 50, 74; on mine sites 77, 79, 117; riding and teams 2, 5, 18, 47, 65; whim turning 22
Hoskins brothers foundry, Sydney (NSW) 63
housing *see* Kalgoorlie: housing at
Huddart Parker shipping company 23
Hunter, Larry 150
Hutton, General Edward T. 63

ice, for cooling in mines 118
Illustrated London News 22
immigrants, to early Kalgoorlie field 3, 11, 23, 46
Industrial Workers of the World 119
industrial relations 113, 118, 119; *see also* trade unions, and unionists
inspectors of mines 118
International Monetary Fund 160
Iranian crisis (1979) 160

Irish-Australians 103
Iron Duke company and mine 16, 124, 144, 151
iron, corrugated, galvanised and flat 25, 107, 128; in cyclone (1928) 126, 127
ironstone 7, 9, 11
Italians, on goldfields 92, 101, 102, 106–7, 112, 141, 143
Ivanhoe Gold Corporation 35, 115
Ivanhoe Gold Exploration 88
Ivanhoe Gold Mining Co. NL 13, 18, 37, 77
Ivanhoe: lode 80; mine 9, 13–14, 29, 30, 50, 89, 92, 96, 108, 130; Patterson shaft 108; reef 30; shaft 84, 130; *see also* Ivanhoe Gold Mining Co. NL *and* Ivanhoe Gold Corporation

jackhammers 94
Jackson, Tom 38
James brothers 44
James, Walter 65
Japanese, on goldfields 47
Johnson, H.V. 144
Jones, Harry C. Rhys 86
Judd, H.A. 88

Kalgoorlie: appearance of early town 25, 65; assayer's shop 32; banks and shops 47, 66, 126; breweries 47; bush scenery 12–13; churches 47, 126; climate and rainfall 3, 5, 13, 30, 68, 69, 70; as commercial centre 47, 48; cyclone (1928) 126; decline after 1974 158, 160; decline in 1920s ch. 7 *passim*; discovery and early goldfield 2–3, 7, 9, 11, 19, 22 *and see also* Hannan's [Find]; electric

INDEX

tramways 48, 66, 169; electricity 103, 130; hospitals 47; hotels 27, 47, 48, 54, 82, 141; housing at 25, 27, 46, 47, 77, 88, 92, 114, 126 (tents, shelters and huts 12, 25); laundries 47; mining boom of 1960s and 1970s 156; population 23, 35, 47, 49, 50, 115, 158, 160; post office and telegraph 18, 23; press 48; recovery in 1930s ch. 8; riot in 1934 141; senior citizens community centre 162; town and private gardens 74; town life staider 92; travel to, and railway 25, 46, 47, 65, 169; *see also* horseracing and racecourses; mines, and mining companies, at Kalgoorlie; water supply

Kalgoorlie Cup 50
Kalgoorlie Electric Power and Lighting Corporation 130, 150
Kalgoorlie Enterprise mine 152
Kalgoorlie Lake View company 158, 159, 165, 166, 167
Kalgoorlie Miner 48, 135, 143
Kalgoorlie Mining Associates 159, 160, 162, 165, 166
Kalgoorlie School of Mines 117, 159
Kalgoorlie Southern Gold Mines NL 152
Kalgoorlie Sun 48
Kalgoorlie Western Argus 48
Kalgu mine 38
Kalgurli company and mine 34, 84, 88
Kalgurli Gold Mines 12
Kallaroo (WA) 25
Kambalda: district 13; mines, town and smelter 156, 162
Kanowna (White Feather) (WA) 32, 47, 56, 58, 76, 115

Kaufman, Charles 44
Kavanagh, Det.-Sgt 83
Keenan, Norbert M. 66, 68
Kirwan, John 48
krennerite (mineral) 34
Krupp Mill 108
Kwinana refinery 156

Labor Party (WA) 97, 139; candidates 119
Lake View & Star company and mine 100, 108, 114–15, 124, 129, 130, 131, 134, 139, 143, 148, 150, 152, 161, 165; ends as company 157–8; power station 130
Lake View and Oroya Exploration 88
Lake View Consols company and mine 9, 14, 30, 35, 46, 50, 82, 84, 88, 92, 94, 100, 101; Duck Pond 35, 82
Lake View Extended mine 38
Lake View South mine 88
Lakewood Main Camp 150
lamps, mine 145
Lancefield mine 129
Lane, Zebina 14
Lane, Zebina B. (son of Zebina) 14, 16, 19, 42, 63
Lawson, Henry 34
lead, acetate of 108
Lennonville (WA) 101
Leonora (WA) 91, 126, 129, 143
Leslie, Bernard 135
Leviathan Quartz Crushing Co. 19
Lindrum, W.A. 54
locomotives, mine 131, 145, 147
London and Hamburg Gold Recovery Co. 44
London, meetings, and control of companies 83, 88, 89, 91, 116, 118, 134, 135, 166
Londonderry mine, Coolgardie 22, 89
Long Reef mine, Lennonville (WA) 101
Long Tunnel mine, Walhalla (Vic.) 29
Lynch, Con 30
Lynch, John J. 135
Lyne, Sir William 66

MacArthur-Forrest *see* cyanide process
MacLaren, Dr Malcolm 115, 130
Maher, John 61
Mahomet, Faiz and Tagh 18
Mahon, Hugh 48
Maier, John Le 38
mail, to London 91
Maldon (Vic.) 82
Manners, J.E. 134, 161
Maritana Gold Mining Co. 14
Maryanski, Modeste 34
Maryborough (Qld) 171
McDermott, Jack 161
McIlwraith, McEachern & Co. 23
McKnight, Jack 38
McLeod, Alec 147
McLeod, Allister 166
Melba, Dame Nellie 63
Melbourne (Vic.) 3, 13, 23, 25, 34, 48, 60, 166
Melbourne Cup 50
Menzies (WA) 25, 32, 47, 54, 61
Menzies, Sir Robert 144, 152
mercury, in gold treatment 40, 42
metallurgy *see* gold: separation and treatment, and treatment plants
Metals Exploration Ltd 163, 166

Mideast Minerals NL 166
mills, Krupp and Griffin 79, 88, 96, 108; *see also* stampers
Mine Workers' Relief Fund 115
Miners' Phthisis Act 116
mines and mining companies, at Kalgoorlie: accidents and safety in 38, 94, 96–7, 131, 148; chimneys 77, 79; decline in war of 1939–45 143; decline of dividends in 1920s 122; decline of goldfield after 1918 ch. 7; decline of mines in 1970s 157, 158–60; dewatering of mines 131, 161; dust and ventilation 94, 96–7, 118, 130, 131; early mining properties 29; exploration and mine recovery from 1979 160–2; headframes (poppet heads) 145, 147, 169; labour and conditions 25, 27, 48, 79, 92, 94, 96–7, 101, 106, 114, 115, 118, 144, 147–8 (*see also* wages); lodes 79–80; machinery 14, 18–19, 42, 77, 145, 147, 148, 168, 169 (*see also* machines by name); mine managers 88–9, 91, 94, 107, 116, 117 (*see also* managers by name); mines flooded (1948) 144; new companies in 1930s 135; new machines and techniques in 1950s 145, 147; open-cut mine (Super Pit) 163, 165, 166, 167–8, 169, 171; opinions of mines and field 18, 19, 22, 29, 34, 77; problems of costs 86, 107–8, 113, 114, 116–17, 128, 144–5, 156; sale of, to London companies 14, 16; shafts 27, 29, 83, 116, 145; strikes 119, 162, (in 1916) 107, (in 1935) 139, effects of woodcutters' 108, 112; surface workings in early 1900s 79; taxation concession, federal 159; underground workings 32, 34, 77, 83, 131, 143, 145, 147, 171; wages paid 114, 118, 119, 128, 139, 144, 157, 162; workers buy farms 119; *see also* gold: output of metal and ore *and* companies by name
Mining Journal, Perth 86
mining laws and leases 5, 7, 9, 13, 16, 47
Mitchell, Deane P. 80, 82
money orders 25
Montgomery, Alex 122
Moonta (SA) 16, 92, 171
Moore, B.H. 117
Moran, C.J. 58
Mount Charlotte: hill 2, 12; mine 152, 156, 157, 160–1, 166, 169; reservoir 66
Mount Coolon mine (Qld) 150
Mount Isa (Qld) 11
Mount Morgan goldfield (Qld) 35, 37
Mount Percy mine: opening ceremony 162–3, 166; treatment plant and open cut 169
mulga 103
Mundaring weir 62, 63, 74, 75
Murchison goldfields (WA) 147, 169
Museum of the Goldfields, Kalgoorlie 147

'nap' (card game) 2
Nepean nickel mine 163
New Kalgurli mine 147
Newcastle (NSW) 50
nickel, discovery, mining and smelting 156, 157, 160, 162
noise, on goldfields 77, 79; *see also* stampers
Normandy Poseidon Ltd 50, 167
Normandy Resources NL 167
Norseman (WA) 13
North Boulder mine 29
North Kalgurli company and mine 124, 136, 139, 143, 156, 157, 159, 162, 163, 166; North Kalgurli (1912) Ltd 134
Northam (WA) 61
Northern Deeps mine 136
nuggets *see* gold

O'Connor, Charles Y.: achievement 68–71, 74–6; death 55, 63, 65; early work in WA 55, 68; in New Zealand 55, 68; pipeline plan 59–62, 68–9, 71; and rainfall question 68–71
O'Connor, Susan L. (wife of C.Y. O'Connor) 63
Observer, Adelaide 13
Oliver, John 161
Olympic Dam (SA) 165
Ord River, pipeline scheme to Perth 68
ore treatment *see* gold: separation and treatment, and treatment plants
Oroya Brownhill, company, mine and lode 77, 80, 82, 83
Oroya Links company and mine 100, 108, 117
Oroya mill, treatment plant 161, 168
Oroya Northern Blocks mine 124

Palace Hotel, Kalgoorlie 48, 54, 83
Paringa mine 124, 143, 152, 163; treatment plant 168
Parkinson, James 8
parliament, of Western Australia 107, 119
Pearce, Samuel W. 6–7, 8, 9, 11–12, 13, 14, 16, 18, 35, 171
Peel River mines (NSW) 89

INDEX

Perseverance mine 16, 50, 84, 88, 94, 108, 124, 152, 161; cyclone (1928) 127; mill 103; poppet head 169; recovery in 1930s 131, 134; royal commission (1905) 91; *see also* Boulder Perseverance
Perth (WA) 1, 5, 7, 25, 38, 47, 50, 55, 59, 60, 62, 74, 162, 166
Phillips, Brian 162
pipeline, water *see* water: goldfields water scheme and pipeline
pneumonia 96
police, on goldfields 82, 83, 113, 141
Port Augusta (SA) 1
Port Phillip mine, Clunes (Vic.) 89
Poseidon Ltd 157, 165, 166, 167
Poseidon, racehorse 50
Postle, Arthur ('The Crimson Flash') 50, 54
powerhouses: steam 116; GMK 156; Lake View & Star 130
Pre-Cambrian rocks 11
Premier, (suction dredge) 60
Price, Julius 22
Pritchard, W.A. 101
prospectors, on Coolgardie and Kalgoorlie fields 2, 5, 6, 8; *see also* Bayley, Arthur; Ford, William; Hannan, Patrick; Flanagan; Shea; Pearce, Samuel W.; Brookman, William G.
prostitutes 47

quartz 2, 7, 11, 12, 23, 34, 96
Queen's Church, Boulder 92

railways: on mine sites and in Kalgoorlie 79; to and from goldfields 23, 25, 86, 103, 115; in WA 62; Trans Australian Railway (transcontinental) and train 107, 118, 126; *see also* Kalgoorlie
Rand goldfields, South Africa 88–9, 117, 129
Ravenswood goldfield (Qld) 40
Reid, Arthur 126
Reid, George H. 66
respirators 96
roasting process and furnaces 40, 42, 44, 46, 96, 108, 169
Robinson mine 58
Robinson, W.S. 150
Rodda, W.H. 88
Roman Catholic miners 92
rope, manilla 107
Ross Creek (Vic.) 23
running, professional 50, 54

salmon gum 103
salt heaps 19
Salvation Army, Kalgoorlie 151
Sammy (Frank Espie's servant) 151
sands, waste 79, 136
Scaddan, John 97, 116, 119
Scots, on goldfields 92
'Scrap Iron Mile' 115
Scullin Labor government 128
Sears Tower, Chicago (US) 84
Sebastopol (Vic.) 23
sharebrokers 32, 34; *see also* stock exchanges, and share trading
Shea, Dan (prospector) 2, 3
shipping and travel to WA 1, 3, 23, 34
silicosis 116
Skewes, Edward 88
'Slavs', on goldfields 101, 106, 141, 143

'slimes' 42, 44, 46, 79, 136
Smith, Howard, shipping company 23
smoke, from ore treatment etc. 46, 77, 79
Snowy Mountains Scheme 65
socialists, on goldfields 97
Sons of Gwalia mine, Leonora (WA) 91, 129, 139, 143
South Australian School of Mines and Industries 37, 148
South Kalgurli company and mine 84, 88, 117
South Kalgurli Consolidated company 134, 143, 152
South Oroya mine 151
Southern Cross (WA) 1, 5; rainfall at 69–70
speculation *see* stock exchanges, and share trading
spitzkasten (classifier) 46
sport, on goldfields 50, 54
stampers, mine and mill 18, 19, 40, 44, 46, 77, 88
Stanton, Mrs, boarding-house 92
steam power 13, 18, 116, 130
steel, shortages of 108, 144
stock exchanges, and share trading: Coolgardie 30; in Kalgoorlie 30, 32, 135; London 14, 18, 29, 32, 37, 44, 48, 135; White Feather 56
Stokes, Ralph 77
Sulman and Teed, cyanide process 46
Sutherland, John 44, 46, 114, 122, 126
Sydney (NSW) 25, 48
sylvanite (mineral) 34

tailings *see* dumps, mine

Teetulpa (SA) 1
telegraph lines 1, 18; telegrams from mines 89
tellurium and telluride ore 32, 34, 40, 42, 144
Thomas, Kingsley, investigates mines (1925) 116–17, 118, 124, 129, 134
Thorn, Joseph F. 129–30, 131, 135, 148; Thorn pump 131
Those Were The Days 126
timber, for underground workings 79, 103
Tonkin, John 157
Toorak, Coolgardie suburb 63
Townsville (Qld) 163
trade unions, and unionists 94, 97, 101, 103, 113, 115, 119, 157, 161–2; *see also* industrial relations
Trafalgar, Kalgoorlie 88
trams: aerial, and buckets 79; mine 100; *see also* Kalgoorlie: electric tramways
Travels in Western Australia 48
trees 13, 22, 25, 27, 46, 74, 103; *see also* firewood
tributers, in mines 108, 119, 122, 124, 128, 131, 134, 135, 143
Triton mine 150
trucks, hand, for ore 131
tube mill 44, 46
tuberculosis 97, 116
'two-up' 128
typhoid fever 86

Vail, H.E. 129
vanadium 80
Vivienne, May 48

Wainwright, W.E. and committee 117
war: of 1914–18, and foreigners on goldfields 106–7; of 1939–45 141, 143, 144; Korean 144
Warrick, John 114
washing, of clothes 13, 47
water schemes and pipeline for goldfields ch.4; artesian water 61, 62, 71; pipeline construction 63; pipeline opening ceremonies 63, 65, 68; plans 55, 56, 58–62, 65; pumping and pumping stations 59, 60, 62, 63, 66; shafts for water 61, 62, 71, 74; supply of plentiful water 86
water: salt water on goldfields 9, 19, 27, 58, 61, 71, 74–5, 117; salt water from Southern Ocean 61; water sprayed in mines 96; supply on early Kalgoorlie field 3, 5, 9, 13, 27, 42, 54; tanks, at mines 79
Wealth of Nations mine 22
Westcott, Lou. 151
Western Mining Corporation 134, 150–1, 152, 156, 158, 159, 165; exploration work 152, 156; withdrawal from Kalgoorlie field 165–6
'Westralia' 19
wet-crushing *see* Diehl, Dr Ludwig

White Feather (WA) *see* Kanowna
Whitfield, C.A. 134
Whitford, Stanley 92, 97
Whitlam federal government 159
Widdop, William 38
Williams, Ernest 134
Williamson, Mrs, house 126
Wilson, J. 58
Wilson, S.R. and W.R. 61
Wiluna mine 129, 130, 139
Winter, A.S. 117
women, on Kalgoorlie field 25, 27, 47, 48, 89, 169; early absence from field 13
Woodall, Roy 156
woodcutters and woodlines 103, 106, 108, 112, 141; *see also* firewood
Woodline 150
Woodward, Harry P. 19, 22
Workers' Compensation Act 116
Wreford, Peter 148
Wright, Whittaker 35

Yates, Colin 147
York (WA) 7, 58
Yugoslavs, on goldfields *see* 'Slavs'